D1119145

A superbly preserved fish from the Jurassic.

THE PRACTICAL PALEONTOLOGIST

Steve Parker

Raymond L. Bernor, Editor

A FIRESIDE BOOK
Published by Simon & Schuster Inc.
New York London Toronto Sydney Tokyo Singapore

A QUARTO BOOK

SIMON AND SCHUSTER/FIRESIDE
Simon & Schuster Building
Rockefeller Center
1230 Avenue of the Americas
New York, New York 10020

First published in Great Britain in 1990
by Quarto Publishing plc

Designed and produced by Quarto Publishing plc,
The Old Brewery, 6 Blundell Street, London N7 9BH

Senior Editor
Kate Kirby
Editors
Phil Wilkinson, Angie Gair, Maggi McCormick
Designer
Neville Graham
Illustrators
John Woodcock, David Kemp
Photography
Simon Powell, Martin Norris
Picture Research
Susan Rose-Smith
Art Director
Moira Clinch
Publishing Director
Janet Slingsby

Typeset by Ampersand Typesetting Ltd
Color reproduction in Hong Kong by Regent Publishing Services Ltd
Printed in China by Leefung-Asco Printers Ltd

3 5 7 9 10 8 6 4
1 3 5 7 9 10 8 6 4 2 Pbk.

Library of Congress Catalog Card Number 90-45178

ISBN: 0-671-69308-5
0-671-69307-7 Pbk.

For information about our audio products, write us at:
Newbridge Book Clubs, 3000 Cindel Drive, Delran, NJ 08370

The author would like to thank William Lindsay, Peter Whybrow,
Paul and Jackie Adams, Dr Mike Benton and Liz Loeffler of the
Department of Geology, Bristol University and staff at Canon Hill Park,
Birmingham Nature Centre.

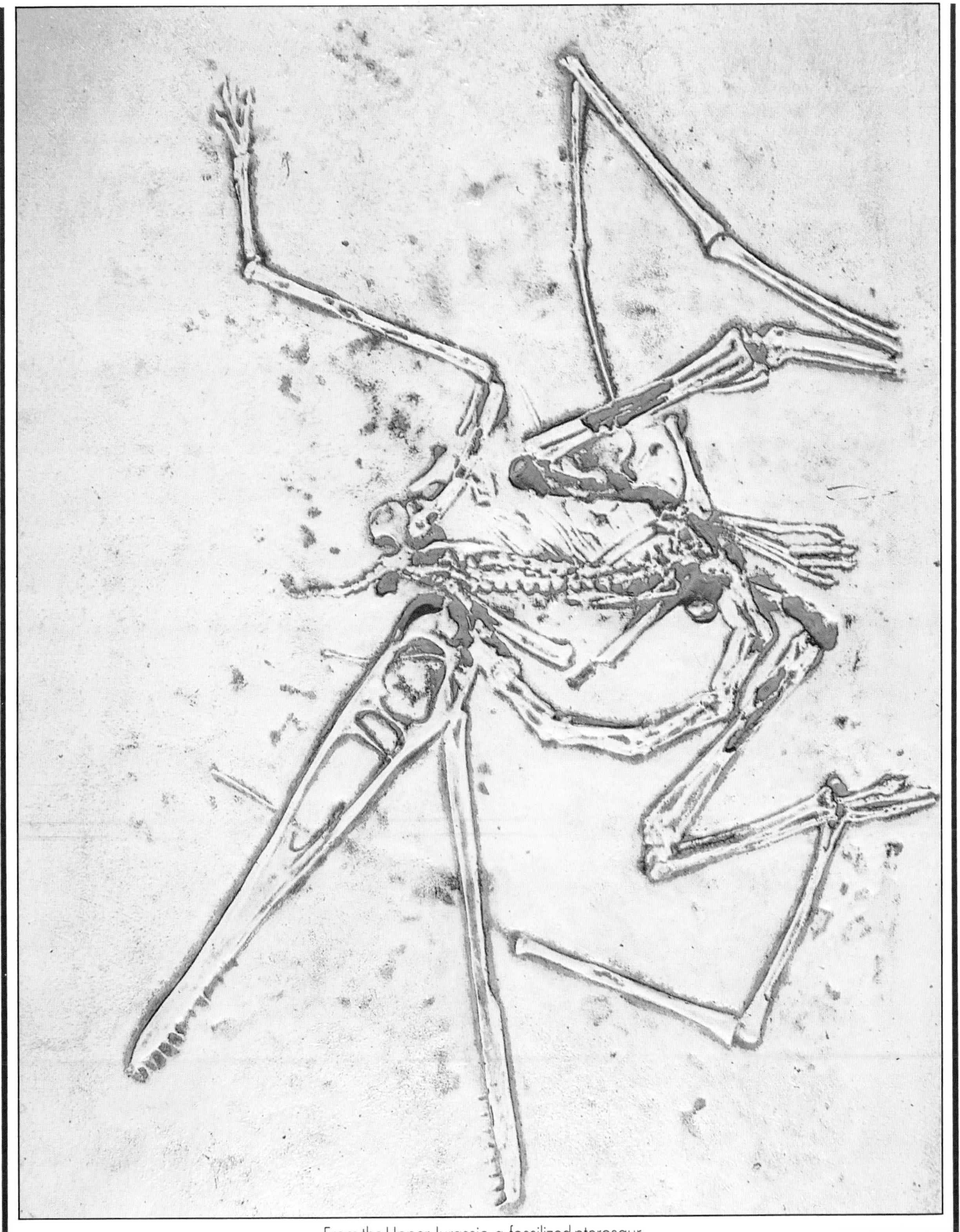

From the Upper Jurassic, a fossilized pterosaur.

CONTENTS

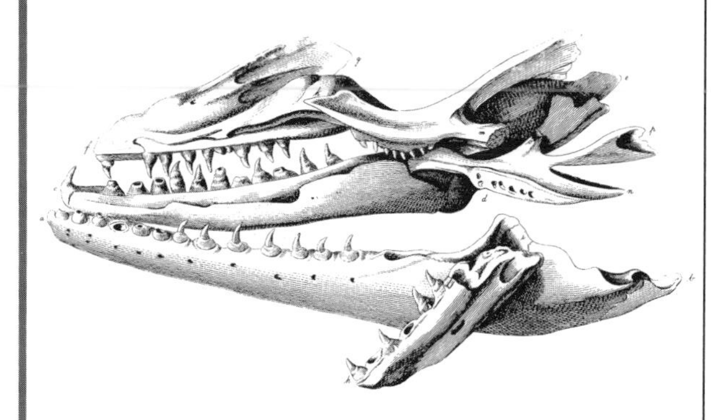

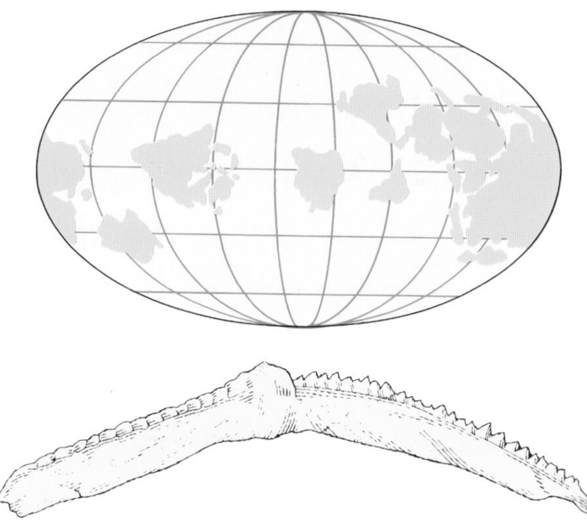

INTRODUCTION

Most people imagine paleontology to be the dry and dusty domain of university professors and museum staff. In fact, these experts are kept informed and supplied by hordes of enthusiastic amateur fossil hunters, who spend weekends hiking and searching quarries and beaches for the smallest fragments of the remains of past life on Earth. Without the efforts of enthusiastic amateurs, museum collections would be far fewer in number, and we would know less about dinosaurs, giant sea-scorpions, the steamy swamps of the coal forests, the origin of life itself – and where our own species came from.

In recent years, fossil hunting has progressively lost its eccentric image to become an active and popular pastime. This is no surprise, for the collection and study of fossils – paleontology – holds many attractions and fascinations.

Something for everyone

There seems to be a basic human need to collect things, be it stamps, antiques, old magazines – or fossils. Some people are attracted to fossil-hunting because they enjoy the outdoors scene, with fresh air and the physical exertion of the search, and the challenges posed by excavating a new find. They may combine their fossil hunting with other interests such as camping, rock-climbing or just traveling.

Others enjoy the delicate work involved in cleaning and preparing their finds. The identification and interpretation of fossils can involve much detective work, gleaning information from popular books and learned scientific works, and this is another fascination for many.

For some, the true satisfaction comes when colleagues admire their finished product. For those who enjoy being by themselves, fossil hunting can be pursued alone in quiet, undisturbed locations. For more sociable individuals, there are often local organizations to join, meetings to attend and interests to share.

Our place in time

The study of rocks and fossils soon brings you in touch with the vastness of time and the great powers of nature. We think of one year as a long time; it is difficult to grasp the fact that the Earth is more than 4,500 million years old. Its immense forces and cycles, the slow yet relentless movements of its crust, the buckling and eroding of mountains, the weather and the tides, and the rise and fall of the oceans, have shaped its past. They will continue to fashion its future.

What we see around us today is a movie "still," a single frame from a very long film. The occasional earth tremors that damage our concrete cities are testimony to the continual shifting of the rocks beneath us.

Life, spawned and nurtured by these irreversible events, has itself become a shaping force over the past 3.9 billion years. The influence of living things on the physical, chemical and climatic cycles of the planet are revealed by fossils. Blue-green algae and simple plants have changed the chemistry of the atmosphere, allowing living things to leave the water and invade the land. Tiny shelled animals have changed the chemistry of the seas, and their fossilized remains are embedded in rocks by the billion.

Natural turnover

About 600 million years ago, when nature invented the hard shell, living forms as most of us would now recognize them began to evolve. Since that time, many more species have appeared, risen to dominate, and then become extinct – far more than the total number of species that exists today.

These species did not die without trace. The traces, as fossil remains preserved in rocks, are the province of paleontology. A few of the animals and plants we see around us bear a resemblance to their ancestors which lived hundreds of millions of years ago. But most life forms have been here for only a short part of the Earth's history. It is like one long experiment.

Human beings are a recent part of the experiment. Our history is measured in millions of years while our destruction of the world's environment is measured in only decades. Study of the past puts into perspective our place in the general scheme of things.

Filling the gaps

As the eminent biologist and geologist Charles Darwin pointed out in his *Origin of Species* (1859), the Earth's history contains more gaps than information. However, fossils show us the broad outlines of evolution and change. Rarely, we can

Sedimentary rocks. Once the bands in these rocks were muddy sediments at the bottom of the sea. Immense pressure has turned them into rocks, buckled and folded them, and thrown them up above the sea – where they are being eroded again, back into sediment at the bottom of the ocean. They hold a record of the animals and plants that lived in the seas above them, as they formed the first time around.

even see the fine details of evolution's workings, when so-called "missing links" are discovered. We can estimate the age of a rock by looking at the fossils it contains, and we can see how rocks have been created, eroded and re-formed.

Throughout the fossil record, there is evidence of mass extinctions. The reasons for these are unclear. Examining the fossil evidence makes us wonder whether today's dominant species will suffer the same fate. What effects will we have on the course of evolution and the future record in the rocks?

How to use this book

The book begins with a section on the history of paleontology. It is a history far shorter than its subject matter, for only in the last century did science realize that the Earth existed long before men and women. Much of the early thinking and research by the great naturalists of the time was directed at reconciling their increasing discoveries and enlightenments with the literal religious teachings prevalent at that period.

Paleontology has since blossomed into a true science. University and museum departments devoted to the subject have devised ever more powerful tools and techniques for analyzing these fragments of ancient stone.

In order to understand the science of paleontology, it is necessary to know something about geology, the study of rocks. The next sections of the book deal with the basic principles that underlie the existence of fossils, including the rock cycle and the forces that drive it, the slow but steady drifting of continents, erosion and sedimentation of rocks, and how an occasional organism has become caught up in the cycle and preserved through the ages. They also deal with the processes of fossilization, the different types of fossils, and the types of rocks most likely to contain them.

This book is intended primarily as a practical guide, both for those who are just realizing their fascination for fossils, and for those who have already encountered the thrill of discovering that a particular rock contained an important find. The following sections offer guidance on planning trips and expeditions, where to look and what to look for, the tools and equipment needed, how to extract

Left Dinosaur bones at Dinosaur National Monument in Utah being carefully excavated from their rock matrix. They will be taken back to the laboratory, cleaned and studied.
Right A fossil crab, *Xanthopsis leachi*, from London clay. This animal lived during the Eocene period, about 40 million years ago. Some creatures have hardly changed since that time.

a fossil from its rock matrix, and how to set up a home workroom and clean and preserve your fossils as part of an attractively presented, well-cataloged collection.

As your interest grows, you may want to specialize in a particular field. Some of the numerous branches of paleontology are described; there are links to crafts and industries, from stone sculpture to the search for fresh reserves of fossil fuels. Some people become immersed in the evolution of a particular group of organisms through the millenia, while others try to recreate how past creatures lived their lives.

There is also information on taking your interest further by consulting journals, visiting and working on important fossil sites, and joining organizations – local or national – where you can learn more.

The history of life

The last section of the book outlines the history of the Earth as we know it, based on the evidence found in rocks and fossils. There are also reference charts on how life has evolved through the ages, and the names geologists have given to various eras and periods in our Earth's history.

From the vaguest beginnings, as microscopic particles which could be the remains of primitive cells, the fossil record progresses to brief and tantalizing glimpses of soft-bodied creatures, then to the explosive evolution of hard-shelled animals that were more easily fossilized. We look at the success of shelled animals, which dominated the seas before the origin of the vertebrates. We see how plants conquered the inhospitable dry land and developed in close association with the insects and other animals that followed them. There are fossils from the great steamy coal forests, which provided the fuels we rely on today. Amphibians were followed by reptiles, including the ever-popular dinosaurs. At the end of the Age of Reptiles, mammals – which had survived for so long in their shadow – soon took over the world.

Finally, we trace the rapid diversification of mammals, culminating in the appearance of our own species, with its "intelligence," agriculture, industry and technology. Somewhere about here, paleontology merges with archaeology, and fossils are found in association with artifacts.

WHAT IS PALEONTOLOGY?

To begin with the word *paleontology* comes from the Greek for "ancient life." This refers to life that is ancient in terms of the history of the planet Earth. In other words, paleontology deals with the life-forms that existed on Earth when its rocks were being created.

Next, the science paleontology has as its aim the scientific study of ancient life. It is by definition an indirect process, because the living things themselves died long ago. The objects of study are the remains which they left on and in the Earth. These remains are called *fossils*, from the Latin for something "dug up from the ground." So paleontology involves the study of fossils, which in turn reveal what ancient life was like.

The word fossil was used originally, in the sixteenth century, for all manner of minerals and metals dug up from the ground. However, it is now applied only to objects obviously created or shaped in the past by living things. As a general guide, such objects which are older than 10,000 years are usually regarded as fossils; younger remains are classed as subfossils.

Paleontology and geology

The surface of the Earth has been changing since the planet was first created some 4.6 billion years ago from a cloud of dust and other matter orbiting the Sun. The changes have been preserved in the rocks that form the Earth's hard outer layer, the crust. Scientists can trace the changes, including the drifting of continents and the uplifting of mountain ranges, by studying the different types of rocks that make up the crust. This is the work of the geologist.

When life appeared on Earth, it too began to leave traces preserved in the rocks. The first life forms were small, soft-bodied and rare; consequently, they left few remains. Only occasionally do we discover their fossils, which give us a precious glimpse into the variety of very early organisms (page 116). In some cases, it is not clear whether a particular rock formation or feature signifies that life was there, or whether it was simply the result of a geological process.

As plants and animals evolved over millions of years, some became larger and more complex. No longer just jelly-like bodies, they developed with hard parts. Their fossilized traces, which are mainly the remains of these hard parts, have been preserved in rocks all over the world. Gradually, such fossils became more distinct and recognizably life-créated, compared to the inorganic, geological features in the rocks. This helps to reassure the lay practical paleontologist who wonders: "Will I recognize a fossil if I see one?" (page 34).

Getting started

If you want to find and study fossils, where do you begin? There are three main starting points. The first is your armchair. Obtain books and magazines about paleontology and fossils. Read and absorb as much as you can, and become familiar with words such as *Jurassic* and *ammonite*. The second is a visit to your local paleontological, geological or archaeological club (ask at the library for details). Attend a few of their meetings and talk to members. This should lead to the third starting point: go on a trip with knowledgeable people, and learn from their experience. Hopefully, you will find yourself a fossil.

The uses of paleontology

A fascinating subject in its own right, paleontology also helps scientists working in other fields. The following are a few examples.

Geology Fossils help geologists to estimate the ages of rocks. They indicate how the surface of the Earth has varied over millennia, and they provide excellent evidence in charting the meanderings of the great land masses.

Climatology Fossils aid paleoclimatologists in their studies of past global weather patterns. This has important bearing on the situation today, when our climate may be irreparably altered by human activities.

Paleoecology From the study of fossils, the history of life on Earth can be written. The environment in which a particular rock was formed, and the kinds of plants and animals preserved as fossils, tell the paleoecologist about habitats in ancient times.

Evolution Paleobiologists look to fossils for information about how life has evolved and who descended from whom, which in turn clarifies the relationships between present-day plants and animals. There is great excitement and public interest when a "missing link" is discovered in the story of life. Paleontology is central to these fields.

The fossil shells of ammonites, mollusks that thrived in the seas for 150 million years, are often found in Mesozoic rocks. Legend explained them as the remains of snakes that had been turned to stone by St. Hilda, in the seventh century. These are, in fact, the remains of animals called *Dactylioceras*, found in Jurassic rocks in Germany.

THE HISTORY OF PALEONTOLOGY

Human curiosity and imagination have always been stimulated by pieces of rock bearing the obvious likenesses of plants and animals. The graves of some Stone-Age peoples contained carefully arranged fossils. Possibly this use had some connection with the beliefs and burial rites of these peoples. But it is unlikely that the origins of the fossils were understood.

Musings of the ancients

The Greek philosophers were the first to record their ideas about fossils. Xenophanes (570-475 B.C.) and Pythagoras (582-500 B.C.) were intrigued, and recognized them as the remains of once-living things that no longer survived in the world – or at least, in ancient Greece. They also proposed that, since fossils of sea creatures were found inland, the sea had once covered the land. Herodotus (485-425 B.C.) specialized in the tiny fossils embedded in the sandstone of the Pyramids. He too realized that an ocean had once covered the Libyan desert, stating that he had identified "shells upon the hills." The great philosopher Aristotle (384-322 B.C.) believed that life was spontaneously generated from mud, and that fossils represented the failures.

For centuries, Chinese people have attributed healing powers to fossils. Traditional pharmacies still sell crushed fossils, such as "dragons' teeth," as cure-alls for everyday ills. However, Chu Hsi, a scholar of the twelfth century, mirrored the Greeks when he deduced that rocks high in the mountains, with fossils of shells in them, had once been seafloor sediments.

"The Devil's work"

The conclusions of the Greeks were forgotten as the teachings of the Christian Bible spread across the Western world. Any scientific view of the origin of fossils would contradict the literal interpretation of the Bible. So, from about the fourth century, the Church taught that fossils had been placed in the ground by the Devil, to tempt people into questioning the truth of the Bible. Those who succumbed

"The animals entered Noah's Ark two by two." Early naturalists suggested that fossils were the remains of those that had "missed the boat." Catastrophes, such as the Flood, were sent so that new, improved creatures could repopulate the world. At each new start, the animals and plants became more like present-day species.

to this temptation, by proposing a non-biblical explanation, would be punished and meet a dreadful end. In general, fossils were to be regarded as "sports of nature." Those which obviously resembled human bones were explained as the remains of the victims – sometimes giants – of Noah's Flood. They were often re-entombed with a formal Christian burial.

Leonardo da Vinci (1452-1519) included fossils in his many studies, and he suspected their actual origin. While investigating layers of fossil-bearing rocks in the Apennines, Leonardo theorized that they had first formed beneath the sea, and somehow later had been uplifted to their present position. As in many spheres, his ideas could have formed the basis of a modern science – in this case, paleontology. But the great artist and inventor recorded his notes in mirror writing and kept them secret from those who would have condemned him as a heretic.

The Principle of Superposition

In 1664, the Danish scholar Niels Steensen, also known as Nicolaus Steno (he latinized his name), traveled to the Mediterranean island of Malta. Here the local traders sold "tongue-stones," which were found all over the island, as good-luck charms. Steno had studied living sharks, and he recognized the charms as sharks' teeth turned to stone. He deduced that Malta must once have been under the sea, and the rocks which contained the teeth had been laid down over a period of time. He further postulated that the deepest rocks were formed first, and were therefore the oldest. This apparently simple, but vital, conclusion is termed the Principle of Superposition, or Steno's Law. However, the scholar later rejected his own interpretation when he turned to the Church and became a bishop.

Despite such setbacks, fossils were quietly and gradually gaining their true status. In Britain, the "Father of Naturalists," John Ray (1627-1705), included rock formations and fossils in his monumental work *The History of the World*.

In the seventeenth century, Niels Steensen studied sharks at the University of Copenhagen. In Malta, he recognized the "tongue-stones," sold as lucky charms, as the teeth of ancient sharks (below). His discovery was not well received by local people, who realized it would jeopardize their tourist trade.

DINOSAURS, DIVERSITY AND EXTINCTION

Some of the most spectacular fossils are those of the dinosaurs. These prehistoric reptiles so intrigued early paleontologists that many of the science's formative years were devoted to their study. The story of how the first dinosaur fossils were found, described and named provides a succinct insight into the development of paleontology as a scientific discipline. It also shows how, as in any branch of science, progress is shaped by the beliefs and personalities of the scientists.

Plot's *Scrotum*

The first dinosaur to be scientifically recorded, illustrated, and named came from the unlikely environs of Oxfordshire, England. In 1676, Robert Plot, keeper of the Ashmolean Museum at Oxford, described and illustrated a fossil which he had been sent. He realized that it was not a chance rock formation, but the end of a huge thigh bone. However, the only animal known to Plot that was large enough to have such a thigh was an elephant. So he assigned the remains to such a creature brought to Britain by the Romans. After further consideration, Plot began to doubt this conclusion, and he finally decided that the bone must

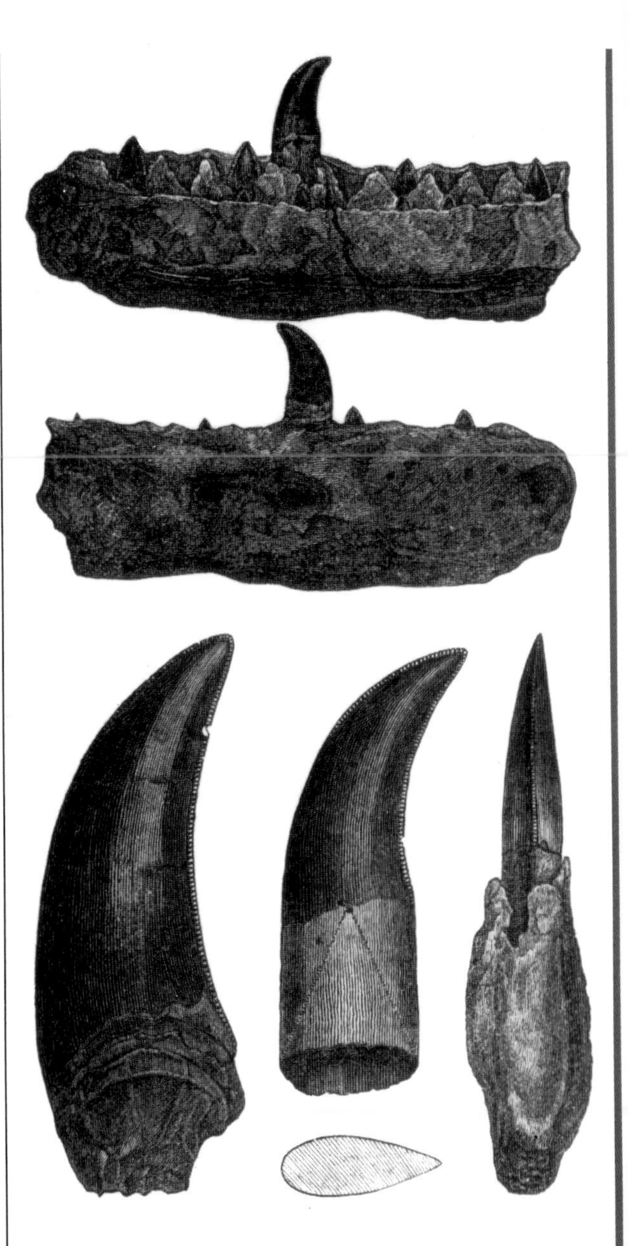

Below Robert Plot studied this fossil carefully. He realized it was part of a thigh bone, but of what animal? A horse or cow was too small. An elephant was too large. Maybe a giant?

Above right William Buckland studied this lower jaw and teeth, also found in Oxfordshire, in the 1820s. He identified the fossils as the remains of a huge meat-eating reptile. He named the animal *Megalosaurus*, or "giant lizard," and published his work in 1836. This animal was the first scientifically named dinosaur.

Right The remains of several similar animals have since been discovered in the Jurassic rocks of Oxfordshire. This is the toe bone of the same creature.

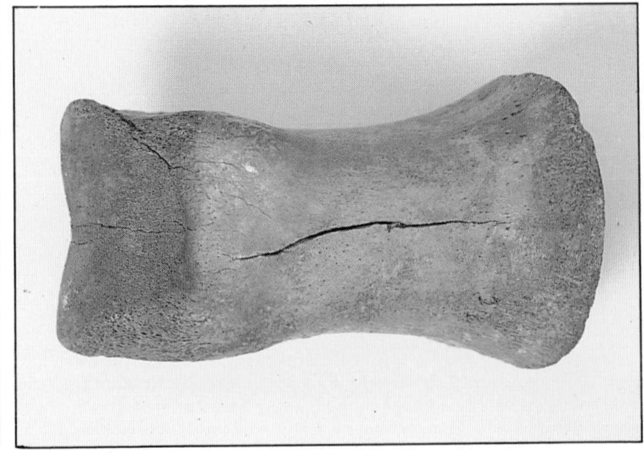

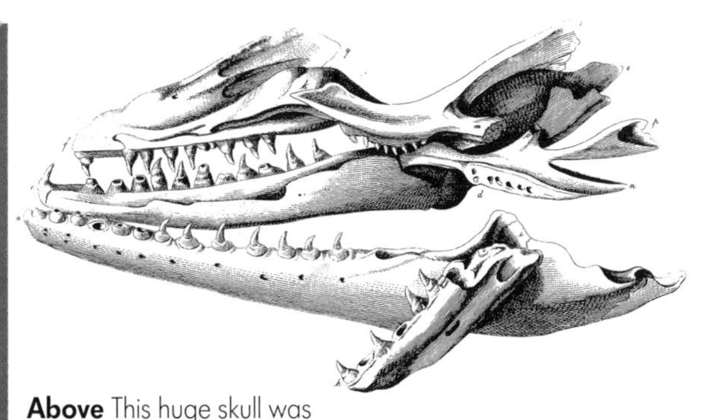

Above This huge skull was found in a chalk quarry at Maastricht, Holland.
Right Baron George Cuvier applied the rules of comparative anatomy to the fossil above, and deduced it was from a marine lizard. Since marine lizards no longer exist, this find fueled his ideas about repeated past extinctions.

have belonged to one of the giants of folklore.

In 1763, the naturalist R. Brookes studied the fossil and came up with a more novel interpretation about its origin. He assumed that some male giant had met a sticky end, and he christened the object *Scrotum humanum*. In fact, this was the first scientific name for a dinosaur, for in 1824 William Buckland, an academic from Oxford University, described and renamed the fossil bone as belonging to *Megalosaurus*, a giant extinct reptile.

During the eighteenth century, many remains of prehistoric mammals such as sloths and mammoths were discovered. Easily excavated from soft, shallow rocks, the skeletons were often almost complete and spectacularly large. Since much of the world was unexplored at that time, people were fascinated by the possibility that such huge monsters might still lurk in the depths of tropical jungles!

The "Father of Paleontology"

In 1780, the fossilized parts of the jaws and skull from a large animal were discovered and excavated from the chalk mines of Maastricht, Holland. To begin with they were thought to be of a whale; then Adrien Camper, a Dutch anatomist, suggested that they belonged to a lizard. During the French Revolution, the remains found their way to Paris, where they were labeled as a crocodile.

Eventually, the foremost expert in comparative anatomy, Baron Cuvier (1769-1832), examined the fossils. He decided, by identifying similarities with animals alive today, that they belonged to a huge sea-dwelling, fish-eating lizard. The creature was later named *Mosasaurus* after its place of origin (*Mosa* in Latin).

Cuvier is now famed as the "Father of Paleontology." He enjoyed the patronage of Napoleon, whom he served first as a naturalist, and later as an official in his administration.

In the late 1780s, the Paris Jardin des Plantes was transformed into a zoo, to house a collection of animals confiscated from their owners by the *gendarmes*. The young Cuvier became fascinated by the animals, especially by the fact that their body structures were often similar, yet different in certain respects. He continued his work at the Paris Museum, where he studied fossils from the Eocene period (55-38 million years ago) which had been unearthed from the plaster-of-Paris mines under Montmartre. Using his knowledge of comparative anatomy, Cuvier was able to reassemble their remains, to make reasoned guesses as to any missing parts, and eventually to reconstruct a view of each animal as it might have appeared in life.

As Cuvier's skills grew, he became the preeminent practical paleontologist of his time, and his fame spread. People sent him fossils from far and wide, hoping that they represented living "monsters" or weird plants which might be brought back at any moment by the many explorers roaming the world. Sadly, this was not to be. It seemed that most of the fossilized creatures and plants had disappeared from the face of the Earth.

The notion of extinction

Cuvier was also interested in geology. On his various trips into quarries and mines, he realized that fossils went deep into the rocks. Those near the surface resembled living forms, but farther down they became less familiar and less recognizable. He also observed that fossils in successive layers of rocks often alternated between land and sea creatures. These findings led him to the conclusion that the organisms whose fossils he held in his hand were no longer represented by living specimens. They had died out completely – in other words, they were extinct.

But Cuvier was a Christian, and struggled to

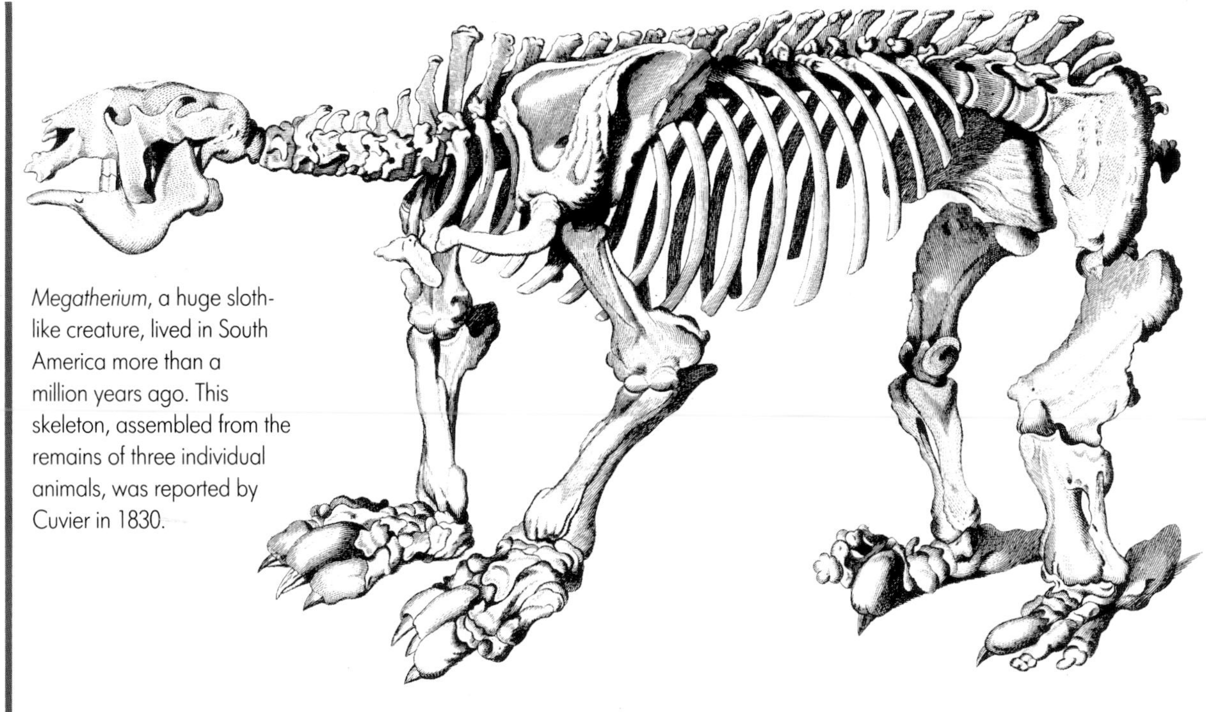

Megatherium, a huge sloth-like creature, lived in South America more than a million years ago. This skeleton, assembled from the remains of three individual animals, was reported by Cuvier in 1830.

reconcile his findings with the Bible's version of the story of creation. His answer was a version of Catastrophe Theory which exists, in various forms, in many cultures. Its basic tenet is that, every so often since the beginning of time, the deity has reassessed the state of the world. Not being entirely pleased, he or she destroyed entire groups of animals and plants by means of catastrophes, such as Noah's Flood. New, improved versions were then created as replacements. Each fresh species continued, unchanged, until it became extinct in the next catastrophe. There was no room in the theory for evolution, with animals and plants changing gradually through time, and giving rise to new species. Cuvier believed that the fossils in the rock layers, or *strata*, were the remains of the destructions. As in Steno's proposal, those nearest the top were the most recent.

Further unearthings

Fossil dinosaur discoveries continued. In 1810, 11-year-old Mary Anning searched and then chipped away the rock on the beach at Lyme Regis, southern England, to reveal a fossil skeleton 30 ft. (10 m) long. It was later designated *Ichthyosaurus* – the "fish-lizard" contemporary with the dinosaurs. Although not a paleontologist herself, Mary Anning grew to be an expert fossil collector who made a

The teeth of the "lizard-fish," *Ichthyosaurus*, preserved in Jurassic rocks. Many of the fossils that Mary Anning collected in southern England were specimens of ichthyosaurs. They were highly prized by fossil collectors of the day.

living by selling "curiosities" from the beach and from a shop in the town, and she later found a *Plesiosaurus* skeleton (although it was many years before her finds were named). The cliffs along the south coast of Britain are still a rich source of fossils, as described later in this book.

Country doctor Gideon Mantell and his wife Mary followed the fashion of the day by being practical paleontologists and keen "fossil hunters." Their book *The Fossils of the South Downs* (1822) contains his descriptions and her drawings, mainly of ancient seashells. Some of the fossil bones and teeth in their collection came from limestones of the Cretaceous period (144-65 million years ago), at a quarry near Tilgate in Sussex. Mantell thought these bones were unusual and set out to identify them. He encountered opposition from the learned people of his day, including Cuvier, who thought they had come from a recent rhinoceros.

Mantell persisted until he met Samuel Stutchbury, at London's Hunterian Museum, who was studying iguana lizards. The two compared the fossil teeth with those of a modern iguana, and decided the former must have belonged to a similar, but gigantic, herbivorous reptile. Mantell reconstructed a creature that walked on all fours, 40 ft. (13 m) long with a spike on its nose, and published his discovery in 1925. *Iguanodon* is now known to have walked on two legs and to have had a spike on its thumb. It was one of the first representatives of the newly discovered group of animals called dinosaurs.

A word of caution: Mantell's passion for collecting fossils eventually ruined both his professional and personal life, and Mary left him.

Gideon Algernon Mantell, 1790-1859. His overwhelming passion for fossils led to the discovery of *Iguanodon*.

Richard Owen, 1804-1892, coined the term *dinosaur*, or "terrible lizard" – although he realized they were not in fact true lizards.

Paleontology comes of age

Richard Owen (1804-1892), a leading anatomist, was Superintendent of London's Natural History Museum. He proposed the term *dinosauria* for the newly discovered "terrible lizards" in 1841, when he delivered a lecture at Plymouth. He believed that the dinosaurs had been created by God to live in far-off times, because they were suited to the climate then. He had scientific ideas about their reptilian cold-bloodedness and benefited in his work from the findings of the biologist and theorist of evolution Charles Darwin. He also held a dinner for leading scientists inside an unfinished model of an *Iguanodon*, on New Year's Eve, 1853.

Darwin – evolutionist and fossil-collector

No history of paleontology is complete without a mention of Charles Darwin, who was a geologist as well as a biologist. He served on the *Beagle's* round-the-world trip in the 1830s, during a decade when the rock strata were being classified and the biblical dates for the Earth's creation were being abandoned. Darwin, while concentrating on geological specimens, also collected fossils of extinct animals to help analyze rock layers – and as evidence for his dawning ideas about evolution. He passed his collection to Richard Owen, who pointed out similarities to living species. The similarities between species separated by time, and also those separated by space (as in the Galapagos Islands), led Darwin to his theory of evolution by natural selection.

Charles Darwin

The main problem with evolution theory was lack of time: it would have taken far longer than biblical estimates of the age of the Earth. Debates continued about the Earth's antiquity, the origins of fossils, and the theory of evolution. But gradually, paleontology took its place in the mainstream of scientific research and progress.

A NEW WORLD OF FOSSILS

The opening up of the "New World" during the past 350 years has led to many exciting fossil finds. Huge, three-toed footprints were discovered early in the nineteenth century, and believed by local farmers to be the footprints of Noah's raven. A collection of assorted teeth from Montana was examined by Joseph Leidy, an anatomist who pioneered the study of North American fossil vertebrates (animals with backbones). He decided they were the first evidence that dinosaurs had lived in North America; in fact, they belonged to "duck-billed" dinosaurs. Leidy also investigated bones found by William Foulke in a marl quarry in Haddonfield, New Jersey. He named the almost complete skeleton *Hadrosaurus foulki*, after the town and the discoverer. This part of North America is now known to be rich in the remains of hadrosaurs, or duck-billed dinosaurs.

Othniel Charles Marsh, 1831-1899. He progressed with great determination through academic paleontology at Yale University.

Edward Drinker Cope, 1840-1897. Talented but with an impetuous streak, he was an independent fossil hunter.

The battle of the fossil hunters

Some of the richest fossil finds to date were uncovered in the American Midwest and West in the 1870s. It was here that amateur paleontologists sparked off a vicious fossil-finding battle. Amateur collector Arthur Lakes communicated his finds to Othniel Charles Marsh, a poor boy who had educated himself using his uncle's money, to become professor of paleontology at Yale University. Marsh did not immediately reply, so Lakes sent some fossil bones to Edward Drinker Cope, a wealthy scientist who had been a student of Leidy's.

Marsh and Cope soon became bitter rivals. The two chased each other around the Midwest during the 1870 and 80s, arguing about who got to each site first and hiding finds from each other. They collected remains from the "badlands" of Colorado, Wyoming, and Montana, and among their discoveries were some of the most famous dinosaurs such as *Brontosaurus* (later renamed *Apatosaurus*), *Diplodocus*, *Stegosaurus* and *Triceratops*. In 1889, Cope ran out of money, and his team stopped digging; but even then the battle continued by publication.

Establishing the tradition

Paleontology itself benefited immeasurably from the frenzied searchings of Cope and Marsh. Many fossil-rich areas were identified, and hundreds of species were discovered. New techniques were invented: Marsh developed the toilet-paper-and-plaster method of protecting bones. Their team members, well trained in practical work, carried on the explorations and established worldwide expertise in the location and study of fossils.

Other names in the list of famous American paleontologists include: Earl Douglas, who found a cache of dinosaur bones in Utah in 1909; Barnum Brown, who collected dinosaur bones from Red Deer Valley in Canada in the late 1890s and early 1900s; Charles Whitney Gilmore, who enlarged the reptile collection of the American Museum of Natural History; Henry Fairfield Osborn, who organized many big fossil expeditions in Colorado, Wyoming and Asia; and Roy Chapman Andrews, of the American Museum of Natural History – discoverer of fossilized dinosaur eggs in the Gobi Desert in the 1920s. More recently, James Jensen has found parts of probably the largest dinosaurs, *Supersaurus* and *Ultrasaurus*, in Colorado during the 1970s; Jack Horner has suggested that dinosaurs nested like birds; and Robert Bakker's suggestions over the past two decades, that the reptilian dinosaurs could have been warm-blooded, have provoked heated discussion.

Left Benjamin Waterhouse Hawkins, a sculptor, was commissioned to sculpt models of the newly-discovered prehistoric "monsters" for exhibition at the Crystal Palace in London, in 1851. He used modern animals as models – plus plenty of imagination. He was very proud of his work, and he and Owen celebrated its completion by holding a New Year's dinner party in the unfinished model of *Iguanodon* (center). *Mastodonsaurus* is left, *Megalosaurus* is right and *Dicynodon* is bottom right.

Above North American Jurassic dinosaurs, *Allosaurus* (left) attacking *Camptosaurus*, displayed in the Los Angeles County Museum. The bones of *Allosaurus* were found by Marsh's field assistants, Benjamin Mudge and M. P. Flech, in Colorado. Many more bones of these animals were discovered in Utah, at the Cleveland-Lloyd Dinosaur Quarry. Study of these bones has revealed the lifestyle of the carnivorous *Allosaurus*. It may have hunted in packs and ambushed the nimble *Camptosaurus*, an iguanodon-like herbivore.

INTO THE 20TH CENTURY

Fossils have now been found on every continent – which means paleontologists were there to uncover them. Werner Janensch, a German paleontologist, discovered dinosaurs during expeditions to what is now Tanzania, in 1909-12. Baron Franz Nopcsa described dinosaurs from Romania during this time. The finds of Scotsman Robert Broom (1866-1951) show that mammals evolved from early mammal-like reptiles, not from later reptiles or birds. In 1947, an Australian, R.C. Sprigg, located Precambrian fossils in South Australia.

The origins of our own species are particularly fascinating. Human evolution has hinged on fossils found mainly in Africa and Asia. Eugène Dubois (1858-1940) obtained the first remains of our immediate ancestor, *Homo erectus*, in 1891. Raymond Arthur Dart from Australia recognized *Australopithecus*, a member of the hominid (prehumans and humans) group, from a quarry near Taung, in Botswana in 1925. The Leakey family – Louis and Mary, and their son Richard – have discovered a great many hominid remains in East Africa during this century.

In the 1980s, there has been an increase in scientific exchange of information with countries such as China and the U.S.S.R., although newly discovered specimens now stay in their original country. Recent expeditions to China have yielded dozens of new species, as well as great insights into the way prehistoric animals lived and the plants that decorated their world. It is another example of how paleontology is still being shaped by the beliefs, attitudes and politics of the day.

A place for the practical amateur

Around the world, thousands of researchers are currently studying hundreds of sites, excavating fossils and transporting them back to the laboratory, where they are reconstructed. Many of the expeditions, or projects, are organized by universities and museums. These trips are expensive to finance, so there must be a very good chance of finding something significant. The workers are often specialized professionals, finding out as much as possible about a limited range of life forms.

The amateur, however, can be broader in scope and wander in less likely territory – and occasionally come up with unexpected "buried treasure."

The claw of *Baryonyx*, which measured 12 in (30 cm) around its outside edge. It could have come from either the front or the hind foot of the animal and was probably used for slashing prey.

Indeed, the larger expeditions depend to a great extent on the finds of amateur fossil hunters to find out where they should invest money in a new project. Such amateur finds are made by luck, or by clever investigation, or more usually by a good slice of both. At localities where new fossils are continually being exposed by erosion of the rocks, the area needs to be searched repeatedly as new strata come to the surface. Amateurs, who can get to know an area well, are unrivaled at this systematic examination.

One exciting find was made by amateur fossil sleuth Bill Walker in 1983, in a quarry in Surrey, England. He found a huge claw in a lump of clay, recognized its importance, and took it to the Natural History Museum in London. An expedition was sent out and recovered about half of the skeleton, and even some fossilized stomach contents, of a previously unknown species of dinosaur. It has been named *Baryonyx walkeri* after its "heavy claw" and its discoverer – an honor that few practical paleontologists would refuse!

Above The Omo River, in southern Ethiopia, cuts through sediments that contain fossils of prehumans and their tools. Many famous prehuman remains have been found in East Africa. In the 1950s and 60s, Louis and Mary Leakey found remains of *Australopithecus* and *Homo habilis*.
Far left A statue celebrates Louis Leakey's contribution to paleoanthropology.
Left Richard Leakey, continuing the family tradition, holds a fossil skull of *Homo erectus*.

WHAT ARE FOSSILS?

Fossils are the preserved remains or traces of living things from long ago. To put these pieces of stone into perspective, and to understand exactly why and how they have formed, it helps to have a background knowledge of the Earth itself and of its age and workings.

The "Age of Earth" debate

The age of the Earth was calculated using the Bible by John Lightfoot, in 1644. The date he arrived at was September 17, 3928 B.C. – at nine o'clock in the morning. James Ussher, Archbishop of Armagh in Ireland, published his calculations in 1650. Using the genealogies and ages of people in the Bible, his estimate for the Creation was October 23, 4004 B.C.

French naturalist Jean-Baptiste Lamarck (1744-1829) realized that the Earth was very much older than biblical estimates. Well before Darwin's time, he proposed that animals and plants had evolved gradually from one form to another. However, he could not supply a convincing reason to explain how – a question that was to be answered by Darwin's theory of evolution by natural selection.

During the nineteenth century, scholars of the rapidly developing sciences of geology and paleontology soon began to realize that their findings did not fit the biblical timescale. The multitude of fossils, contained in rock layers thousands of feet deep, could not have accumulated in only 6,000 years – assuming that present-day geological processes such as weathering and erosion happened in the past as they do today (page 26).

Uniformitarianism and stratigraphy

The Scottish geologist James Hutton (1726-97) began his working life as a farmer and law clerk. He moved on to study the rocks of Scotland. He deduced that many were formed from still older rocks, by a process of gradual change. Given enough time – far longer than 6,000 years – all present-day geological features could have been created in the past by repeated cycles of processes still happening today: erosion, deposition, burial and uplift. Hutton decided to put no time limit on the Earth: it had no beginning. In 1830, English geologist Charles Lyell published a version of Hutton's theory, called the Uniformitarian Principle, in his *Principles of Geology*. It proposed that the Earth was shaped in the past by processes still operating today; the theory became a cornerstone of modern geology and paleontology.

William Smith (1769-1839) was an English mine and canal surveyor. His work brought him into contact with many fossils, and wherever he went he noticed that similar types of rock always held the same types of fossil. Linking this to the idea that the oldest rock strata were deepest (unless they had been disturbed), Smith developed the idea of *stratigraphy*. He produced tables of fossils to be found in certain rocks, and he pinpointed "indicator species" (page 36). It therefore became possible to date each rock stratum, relative to the next. Stratigraphic columns are important tools in geology and paleontology, as are the geological maps (page 42) of the kind that Smith made wherever his work took him.

Other scientists became involved in the study of stratigraphy, and the layers of rocks were gradually classified and named. For example, the Triassic period (225-193 million years ago) was named in 1824 from a threefold division of rocks in Germany. The Jurassic period (193-136 million years ago), the era of the dinosaurs, took its title from the Jura Mountains of the European Alps.

Absolute dating

Relative dating gives the relative age of a rock, based on the fossils it contains and the rocks and fossils adjacent to it. But for many years, there was

James Hutton, 1726-1797. He tried several professions in his early years, before settling down as a "gentleman of leisure" to enjoy the study of geology in his last 30 years.

William Smith, 1769-1839. A poorly educated man, he taught himself the science of surveying. His principle of stratigraphy seems obvious today, but was a major advance at the time.

no clear idea of the absolute age of each rock layer, or of the Earth itself.

In 1896, the French physicist Antoine Becquerel discovered radioactivity. The decay of atoms in unstable radioactive substances, such as uranium, gradually produces stable non-radioactive substances, like lead. The decay happens at a constant rate: the "half-life" of a substance is the time taken for half of a given quantity of it to decay to its stable form. Because all rocks contain unstable radioactive substances (albeit in minute amounts), the ratio between the unstable versions and their stable derivatives can be measured. This leads to a fairly accurate estimate of the time since the rock was formed.

Using this method, the oldest rocks found on Earth, from northern Canada, are 3,960 million years old. Other rocks have also been dated, as charted on pages 110-111. The ages of meteorites and Moon rocks have also been measured by radiometric dating, at 4,600 million years. This is close to the age of our Earth.

The Grand Canyon, Arizona. The Colorado River has cut down to sedimentary rock beds 2 billion years old. Besides creating a spectacular landscape, layers provide a record of the passage of time, and of the creatures that once lived in the warm, shallow seas which covered the area.

What's in a name?

The names of geological periods in history have often been derived from the types of rocks formed at the time, or from places where these rocks are found.

Cambrian From *Cambria*, the Roman name for Wales, for sandstones and shales described from North Wales.

Ordovician From *Ordovices*, a Celtic people who lived in northwestern Wales, where the rocks were first studied.

Silurian *Silures*, another people from central Wales, gave their name to the shales, sandstones and limestones first described there.

Devonian Rocks from this period were first studied in Devonshire, England.

Cretaceous From the Latin *creta* for "chalk," referring to the enormous thicknesses of chalk from this period.

DRIFTING CONTINENTS

The surface of the Earth is never still. Although its movements are virtually unnoticeable, the continents drift as the seas widen, and some mountains rise to the sky while others erode. The continents float on huge blocks of rock, called tectonic plates, that move around the Earth's crust like the pieces of a giant spherical jigsaw puzzle. Thus the continents split, drift apart and then coalesce to form one vast "supercontinent." Scientists speculate that the cycle has taken some 440 million years, and has happened several times in the past.

The global movements are recorded in the rocks. The jigsaw-like fit of continents, such as that of western Africa with eastern South America, was first noticed by the writer Francis Bacon in 1620. The theory of Continental Drift was put forward by F. B. Taylor in America, and by the German Alfred Wegener early in the twentieth century. By studying types of rocks, and the fossils found in them, continental positions can be traced back for over 200 million years.

Pangaea is the name given to the supercontinent which formed 300 million years ago. It split into two main land masses, Laurasia and Gondwanaland, 200 million years ago. They then drew apart, at a rate of a few inches each year, as tectonic plates carried away the land masses that were to become North America, Eurasia, Africa, South America, India, Antarctica and Australia. The movements are driven by the gigantic forces of heat and pressure beneath the plates, termed convection currents. They caused sea levels to rise and fall dramatically, which in turn had drastic effects on life – as revealed by fossil records.

The rock cycle

Superimposed on this global tectonic activity, and driven by it, is the geological cycle: erosion, sedimentation, burial, metamorphism (that is, heating and compaction) and uplift. A *rock* is usually defined as a solid mass of fused minerals found in the Earth's crust. Its ultimate origin is magma, the molten material beneath the crust. Magma is a complex mixture of chemicals in solid, liquid and gaseous forms. It rises to the surface at volcanoes and between the joints of tectonic plates beneath the oceans. Here it cools and solidifies into one of the three main rock types: *igneous* rocks, such as basalt and granite.

Greenland – the rock cycle in action. Sedimentary rocks, laid down in Precambrian times, have been thrown up by the buckling of the Earth's surface. A glacier has worn the surface smooth.

Rocks at the surface of the Earth are subject to erosion. The toughest and hardest erode the slowest, but eventually they are all broken down into small particles by *erosion* – the action of heat, cold, wind, water and chemicals.

These particles are transported by wind and water, to settle as sediments. This process occurs in deserts and river deltas, beneath glaciers and on the ocean floor. Rivers carry the rock particles, known as alluvial deposits, to the sea. The smallest particles float best and are carried far out into the ocean. Coarser particles sink earlier and are laid down nearer the shore, as sand, silt or mud, according to their size. This is the *sedimentation* stage (see opposite).

As the sediments build up and become thicker, the particles are squeezed and compacted by the weight of their gathering colleagues above. Eventually, they are cemented together – *consolidated* – by physical and chemical processes, to form *sedimentary* rocks. This is the second main rock type, and includes sandstones, limestones and shales.

If the rock becomes very deeply buried, it may undergo changes caused by extreme heat and pressure, to become the third main rock type: *metamorphic* rock. Slates and schists are examples.

Any rock may eventually be returned to the molten magma beneath the crust. Or any rock, igneous or sedimentary or metamorphic, may be thrown up by the folding of the Earth's surface, as the tectonic plates jostle together and buckle. *Uplift* has occurred, and erosion begins the rock cycle again.

The rock cycle

The rock cycle has been continuing from the beginning of time – since the Earth first formed. It has no beginning nor end. Some rocks made in the earliest times have lain unchanged for almost 4 billion years. Others contain minerals that have been through the cycle possibly hundreds of times. The rock cycle is part of a larger natural scheme involving the cycles of water, carbon and many other naturally occurring substances.

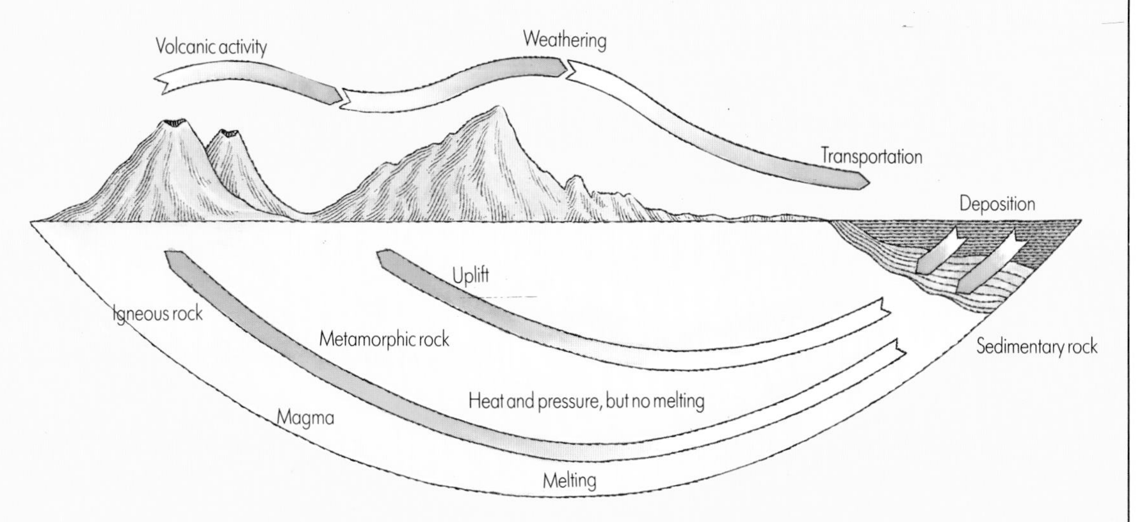

Glossopteris, a Permian plant, grew on Pangaea. Its fossils are found today on the continents that once were part of that supercontinent.

Metamorphic rocks, like this schist, are formed by high pressures and temperatures acting on other rocks. Fossils found in them are distorted or destroyed.

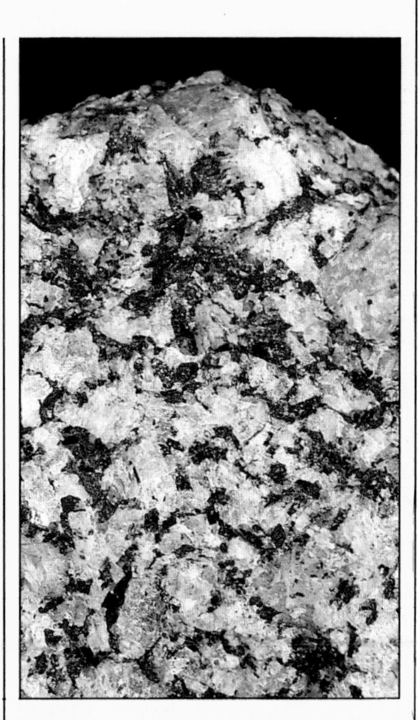

Gabbro is an igneous rock derived from the magma beneath the Earth's crust. Texture and mineral content are influenced by the conditions under which it solidified.

SEDIMENTS AND STRATA

The immense temperatures and pressures which create metamorphic rocks also destroy any fossils they might contain. So does the melting and solidifying that forms igneous rock. The rocks of interest to the paleontologist are therefore sedimentary rocks. Plants and animals destined for preservation are trapped in the process and become buried by the particles laid down during sedimentation. As the conditions of erosion change, with altering climates and land movements, the sediments take on a new character. These changes can be seen at the *bedding planes* that separate distinct layers of deposits. Bedding planes mark out the march of time by delimiting stratigraphic groupings or systems.

Above The eroded hills of Cape Province, South Africa, show the fossil-rich beds formed from sediments about 250 million years old.

Below Variations in this sandstone occurred as the underwater sand dunes from which it formed were moved along by the water's currents.

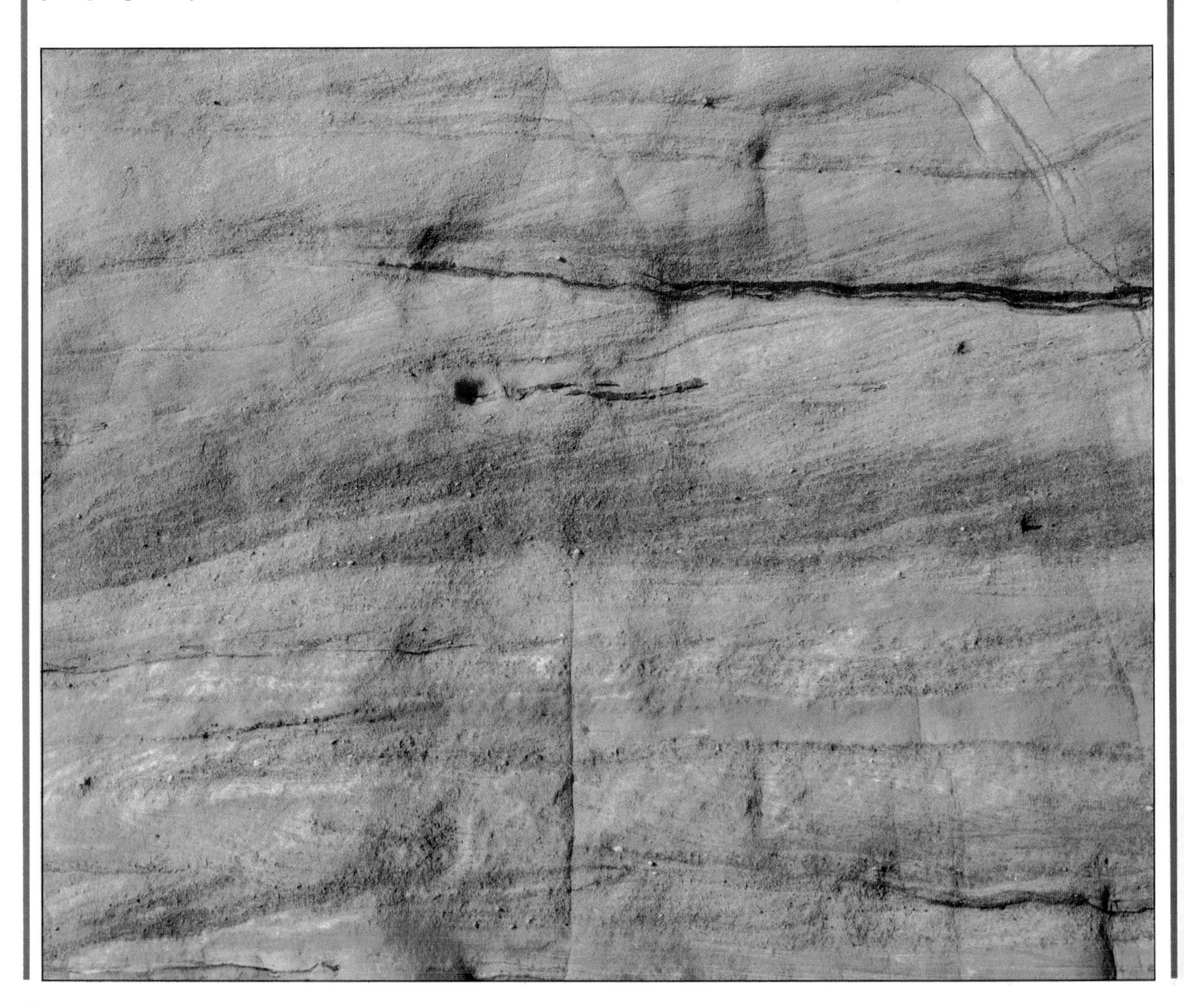

Sedimentary rocks are classified according to the origin and size of their particles. The conditions under which sediments accumulate vary so much that there are numerous classifications, which overlap to give hundreds of sedimentary rock types. In addition, rocks often consist of mixtures of other rocks. The ultimate classification consists of the individual rock beds found at particular sites, with each named for the place where it occurs, as well as for its predominant rock type.

Evaporites

Sedimentary rocks are of three main types. One type is formed from chemically derived sediments. For example, an inland sea might evaporate and leave a deposit of rock salt, or a silaceous (silicon-based) deposit may form in a hot spring. These rocks are described as *evaporites*.

Fossiliferous sediments

The second type is derived entirely from organic material – and is therefore a "fossil" in its own right! Some of these so-called *fossiliferous* rocks change very little from their original state, while others may be deeply buried and change dramatically, due to the great pressure. Many limestones and chalks are formed from the calcareous (calcium-based) skeletons of tiny organisms deposited on the seabed. Some limestones are fossilized corals; others, known as tufa, are derived from mosses and other plants that grow beside hot springs. Carboniferous (carbon-based) rocks such as coal and jet are the remains of plant material laid down in huge quantities; ironstone is produced by algal body chemistry; and coprolites are the fossilized feces of land animals, found in such large quantities in some places that they have been

Right Fossiliferous sedimentary rock formed from the shells of bryozoans, corals, brachiopods, crinoids and other animals. They fell to the bottom of a coral reef face when they died. The shells, and the rock they form, are mainly calcium, and are virtually unchanged from their original state. This formation, the Dudley Assemblage, was shaped in the Silurian period.

Left Cretaceous chalks are hundreds of meters thick in some places. Here it is visible as high cliffs on the south coast of England. Chalk, a very pure form of limestone, is a fossiliferous sedimentary rock. It was formed from the shells of microscopic planktonic organisms that fell to the bottom of the sea. The resulting mud became cemented by minerals and compacted to form a solid rock, under the weight of the sediments above.

used for centuries as fertilizer. The silaceous remains of sponges and microscopic diatoms constitute rocks such as chert and flint.

Clastics

The third type of sedimentary rock is *clastic*. It is formed from eroded particles of other rocks and graded according to the size of these particles. Breccia has coarse, angular, unweathered grains cemented together; it probably represents the sudden collapse of a rock face or cliff. Conglomerate is made of coarse but rounded pebbles cemented together; it is in effect a solidified pebble beach. Gritstone, sandstone and mudstone reflect decreasing particle size. Not all sand is found on beaches, and sandstones may be of desert or river origin. Very fine clays form argillaceous deposits; shales were deposited in riverbeds or on the sea

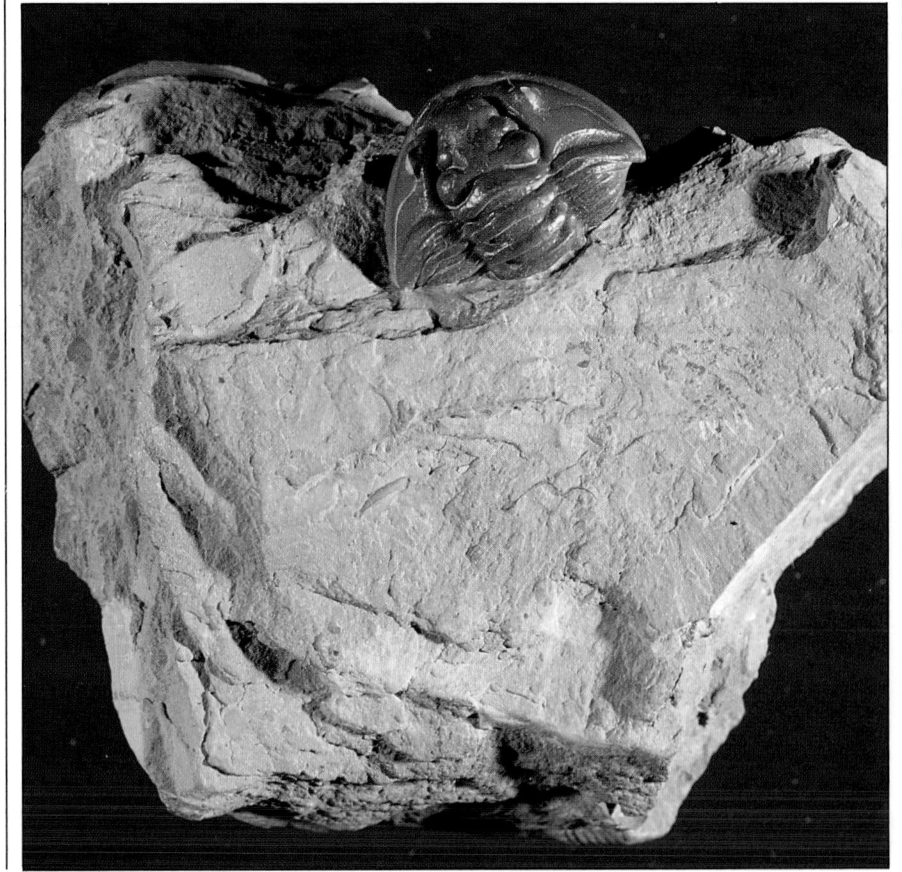

Right Sediments of erosion products are sometimes laid down in places other than under water. This red desert sandstone consists of grains worn into smooth spheres as they were bounced along by the wind. The color comes from the iron oxide that binds the particles together.
Below right This Lower Palaeozoic shale contains a specimen of *Flexicalymene retorsa*, a trilobite that could roll itself into a ball. Shales are fine-grained, sea-deposited rocks made from consolidated mud.
Below Fine sediments laid down as a pavement deposit in the Natal region of Southern Africa.

Gaps in the fossil record

The Earth is in such a dynamic state that nothing remains the same for long, geologically speaking. Climates change, rivers and glaciers alter course, seas rise and fall. Animals and plants flourish and then die to make way for new forms. The sediments that are laid down also change. The result is layers, or strata, of different sorts and thicknesses of rock, one on top of the other, each containing different kinds of fossils. Between the layers are bedding planes. The youngest layer is at the top – except where movements of the crust have tilted or even inverted the strata.

So, sediments have built up, layer upon layer, bearing a sequence of fossils which change with time. Yet it is far from being a neat, continuous story. There are often gaps in the record, where sedimentary rocks are missing from the sequence. Such a gap is known as an *unconformity*. Charles Darwin suggested that more time is unrepresented by unconformities than is represented by sediments. This seems to be true on the land masses, where uplift, folding and erosion of the rocks has carved up and thrown away large chunks of the fossil record. But deep boreholes in the seabed reveal unbroken fossil deposits stretching back 200 million years.

The layering of the rocks is shown to spectacular effect in the Grand Canyon. The limestone rocks near the top are 200 million years old and bear the fossils of reptiles, insects and ferns. About halfway down the cliffs, rocks 400 million years old contain early types of armored fish. On the canyon's floor, in rocks 2 billion years old, signs of life are absent.

Sediments are always laid down horizontally during their original formation. Unconformities occur where the pattern of the layers suddenly changes; they represent missing rocks. Angular unconformities occur where the layers are at noticeably differing angles. These two diagrams show actual examples, from Scotland (above) and Washington state (below).

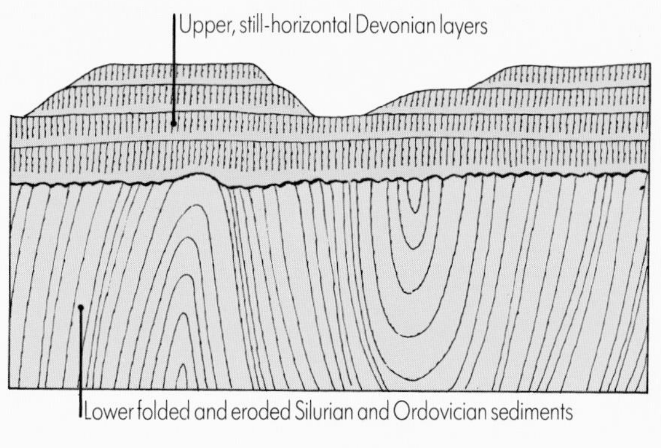

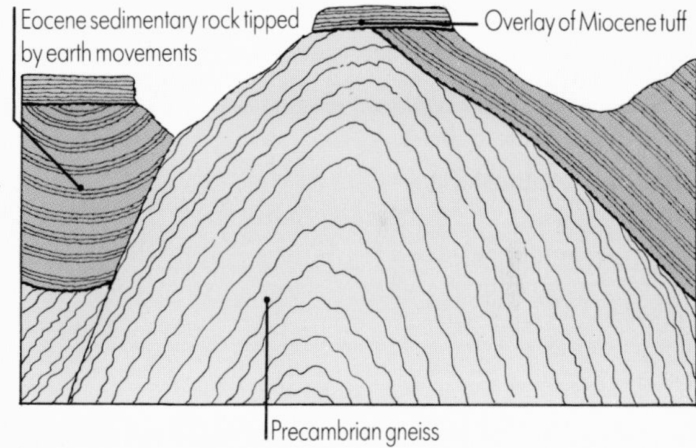

floor. Fine shales are perhaps the most significant sedimentary rocks covering the Earth.

The most common sedimentary rocks

There are many more types and subtypes of sedimentary rocks than those described above, but about 99 percent fall into the three main categories. Because most of these sediments are believed to lie at the bottom of the oceans, it is difficult to calculate the proportions of each. However, it is estimated that approximately 20 percent of sedimentary rocks are limestones, 30 percent are sandstones and 50 percent are shales. These rocks certainly provide enough material to keep paleontologists busy for many years to come.

Likely hunting grounds

The sedimentary rocks most likely to contain fossils are those that were laid down in places where there was abundant life, and where deposition was rapid enough to bury the organisms before their bodies were broken up and rotted away. The sandy bottoms of shallow calm seas, river deltas, lagoons and deserts are the most likely places, so the sandstones, mudstones and shales are good candidates. The finer the sediment, the finer the detail recorded in the fossil. Details such as the feathers of the earliest bird, *Archaeopteryx*, or the fur of those reptilian flyers, the pterosaurs, are only visible because the animals were fossilized in exceptionally fine limestone.

THE PROCESS OF FOSSILIZATION

Finding fossils has been likened to finding "squashed insects between the pages of a book millions of years old." Fossilization is a very rare occurrence, because natural processes tend to recycle as much as possible. Almost all animals and plants that die are eaten or decompose. Even hard parts such as wood, shells and bones eventually break down; and their nutrients are recycled by the action of carrion-feeders, insects, molds and bacteria. The Sun, the air we breathe and chemicals in the soil also help the gradual processes of breakdown and decomposition.

So it follows that fossils are most likely to be preserved where the conditions for decomposition are poorest: where moisture, oxygen or warmth are excluded, or where there are lethal toxins or extreme heat or pressure. The bottom of the sea is usually a good bet, as animals and plant remains sink into a stagnant ooze. Land organisms are only likely to fossilize if they are buried in hot, dry sand or if their bodies lodge in river mouths or inland waters to be buried rapidly by mud, or if they are covered by volcanic ash from an eruption or fall into a bubbling pit of natural tar.

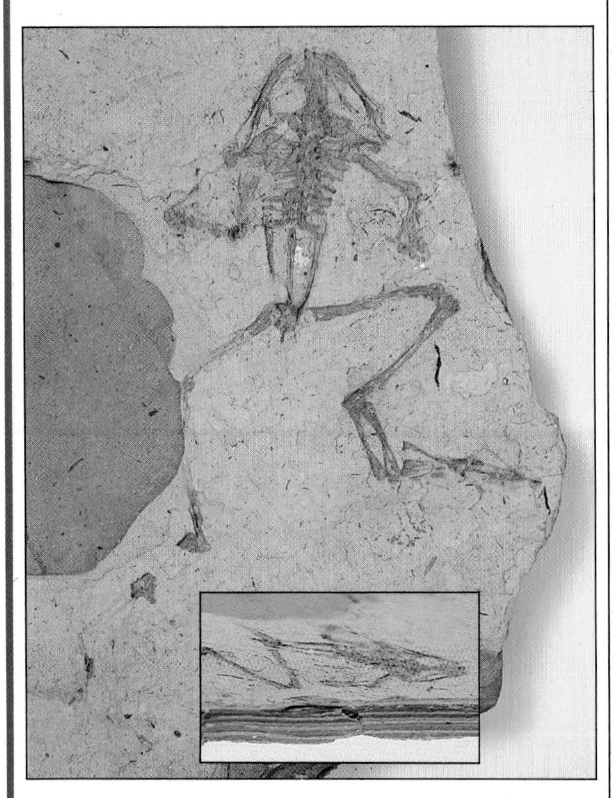

Fossils of animals as fragile as this frog, from Spain, are very rare. Frogs appeared 170 million years ago; only those whose bodies were quickly covered by fine sediment are as well preserved as this. The inset shows how the frog was flattened between the narrow bedding planes of this Pliocene rock.

Permineralization

After burial, the sediment containing the remains gradually sinks as more sediments collect on top. Ever so slowly, it turns to rock. Within this forming rock, very hard remains such as teeth may retain their original state for hundreds of thousands of years. But less resistant parts slowly change. The organic materials such as proteins, hallmark of living origins, disappear. The spaces remaining fill with minerals which precipitate from water seeping through the rock. The process is called permineralization.

Some of the minerals contained in the original organism, such as the calcium in bones, may be replaced by new minerals brought in by the water. The fossil that forms is *petrified*, literally "turned to stone."

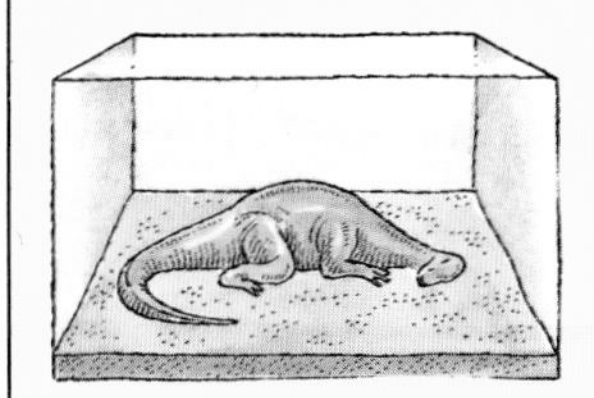

1 The animal dies and falls to the sea or lake bed. The soft parts are decomposed by bacteria, but the hard parts remain.

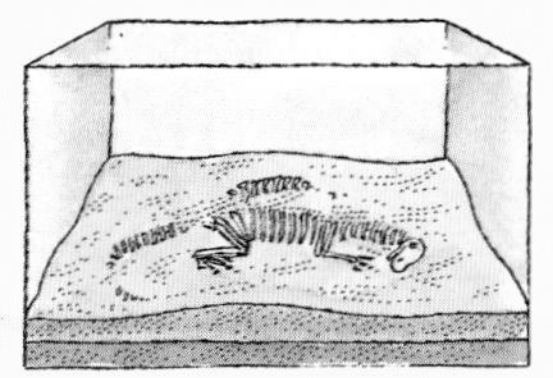

2 Sediments continually rain down from the water above and settle on and around the skeleton.

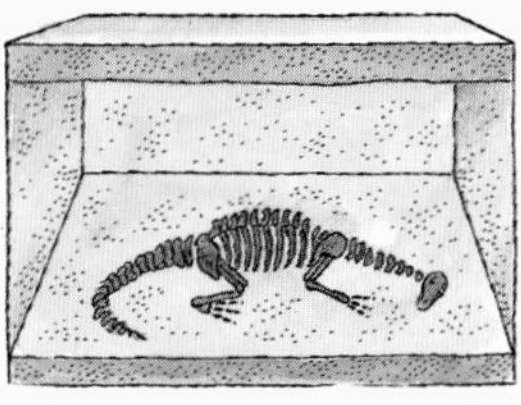

3 The sediments compact into rock, and percolating water slowly replaces the chemicals in the bones with hard minerals.

4 Percolating water may dissolve the bones, leaving a mold fossil, or fill the mold with minerals, forming a cast fossil.

Mold, cast and trace fossils

Percolating water dissolves away the remains of the buried organism completely. This process leaves a hole of the same overall shape as the remains, which is termed a *mold fossil*.

As a further option, the mold may then fill with yet more minerals. The result is a petrified *cast fossil*, though of course it does not contain the fine internal structure of the original remains.

Very rarely, rapid burial in ideal conditions allows the preservation of soft tissues. They may just be impressions in the sand, like the ridges made by a stranded jellyfish on the beach. Or they may be a layer of black, oily carbon left when all the other chemicals that make up the flesh have decomposed and disappeared.

Trace fossils are preserved parts or signs left behind as animals went about their lives. They include burrows, tracks, footprints, eggs and shells, nests and feces.

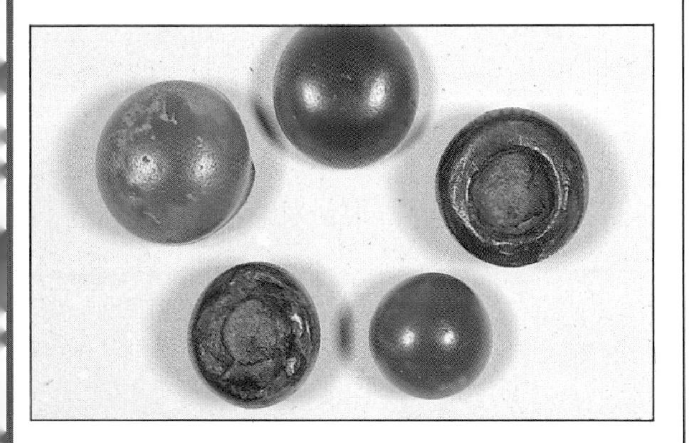

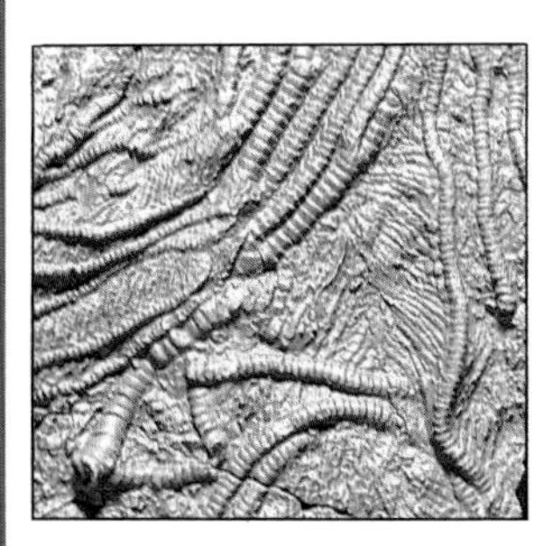

Above The hard parts of animals fossilize best. These fish vertebrae, or backbones, have survived, while the rest of the animal has been lost. **Left** The skins of sea lilies, or crinoids, contain chalky plates that fossilize well. These are from the Jurassic.

Trapped for posterity

Some of the most perfect fossils are preserved in tar, ice or resin. Insects attracted to the sweetness of resin ("sap") oozing from damaged trees were caught by its stickiness. Lumps of resin have been buried in sandstones or mudstones and turned to solid amber. One particularly rich site is La Toca, in the Dominican Republic. Miners in the 40-million-year-old mudstones come upon insects, spiders and even small lizards, perfect in every detail.

Which parts fossilize?

Most fossils of vertebrate animals are of their hardest parts – teeth. In fact, much work in vertebrate evolution is based on studies of teeth. Bones also fossilize well. Perhaps the most common fossils are the shells of aquatic creatures, which were usually durable and already in a situation suitable for fossilization. Typical plant fossils include wood and bark, leaf veins and the tough outer casings of seeds, spores and tiny pollen grains.

Far left Trilobites, like other arthropods such as crabs, lost their shells as they grew. This "mold fossil" was formed 500 million years ago, when the sediments solidified around an outgrown shell, which then disappeared. **Left** Two hundred million years ago a three-toed dinosaur strolled across a muddy shore. The footprints were rapidly filled with sediment, which preserved them to this day as "trace fossils." Many such tracks have been found.

HOW TO RECOGNIZE A FOSSIL

When looking for fossils, there is no substitute for the experience of recognizing biological shapes. Your eye will be caught by likely fossils – once you have seen and identified the first few. Most of us know the sensation of looking hard but not seeing, until that initial find; then we notice that whatever we are searching for is all around.

With practice, you should begin to notice regular patterns and shapes in the rocks, which are more likely to have living rather than non-living origins. This is the key to successful fossil hunting. Books can help, but the fossils themselves are out in the field, and they will not make their own way to the collector's case.

Pseudofossils

There are regular patterns in the rocks that are not produced by living processes, which can fool the collector. They are called *pseudofossils*, and even the experts are sometimes fooled. Controversy rages continuously, particularly over supposed evidence of very early forms of life, which are generally small and undistinguished.

The earliest fossils were long thought to be stromatolites, concentric rings formed by blue-green algae, and found in the most ancient rocks in Canada. Ennobled with the name *Eozoon* ("dawn animal"), some of these "fossils" are in fact layers of calcite and serpentine minerals, probably of volcanic origin. Other false fossils include strange hexagonal mesh-markings in some very old rocks, and the beautiful branching patterns of "moss agates" which look so much like plants. These forms are produced by the percolation of chemicals through fissures in the rocks. Mineral crystals usually have straight edges, but iron ore sometimes forms into round shapes known descriptively as kidney ore. Flint nodules are also fashioned by chance into "animal" shapes and are celebrated in the gardens and walls of old houses.

At the back of your mind, be aware of modern artefacts which look old. One of the early supporters of the idea of fossilization, German professor Jean-Barthelemy Beringer, was fooled by pranksters. He was ridiculed when it was discovered that some of his most interesting fossils – of mating frogs, birds, reptiles and insects – were in fact made of terracotta. They had been hidden in the quarries that he searched by overzealous students.

Climacograptus in Ordovician rock. Graptolite fossils could be mistaken for scratches on the gray shales in which they are often found. They are the remains of the skeletons, or rhabdosomes, of a colony of tiny worm-like creatures. In life, this one looked like a flattened cylindrical version of a gardener's "strawberry tower," with a cup under each hole where the animals lived.

Above *Monograptus* is one of the most common graptolites found in rocks, and occurs throughout the geographical and geological range of graptolites. These animals floated in the plankton of the seas. They were preserved when they fell to the bottom, where the water did not contain enough oxygen for decomposition.
Left This fossil, *Chondrites*, is not the animal itself, but the tunnel in which it lived. These trace fossils occur throughout Paleozoic rocks, but we have no details about the creatures that made them. The branching tunnels were bored in the mud of the sea-bed, possibly by animals that fed on sediment.

INDEX FOSSILS

The early paleontologists soon noticed patterns in the fossil record: ammonites were often found with reptiles, or corals with certain kinds of oysters. Fossils that are abundant, but distinctive, and that evolved fairly gradually in a well-defined series, can be used to date the rocks. They are termed *index fossils*, and the shells of invertebrate animals are often ideal. But they have their limitations. For example, the original animal's habitat restricts the number of places where it may be found.

Many types of mollusks, as well as brachiopods and arthropods make good indicator fossils. Their evolution can be traced in detail back through the rock strata. Smaller creatures that are useful include graptolites, shelled colonies of tiny organisms which have left no descendants. They are found in very ancient rocks, and they gradually diversified into a wide variety of shapes.

Microfossils

Tiny fossils visible only under the microscope, such as conodonts, are known as microfossils. They make good index species. Conodonts look like tiny teeth, and they can be extracted from rocks using dilute acid (page 73). They crop up in older rocks over a very long timespan, and their various shapes make them ideal indexes. The owners of the teeth remained a mystery until 1983, when the fossil of a small soft-bodied animal with conodonts around its head region was found near Edinburgh, Scotland. The animal resembled the modern-day lancelet or amphioxus, a link between invertebrates and the most primitive fish.

Microscopic examination reveals a host of tiny fossils used to indicate the age and type of rock brought up by drilling rigs. Shelled, single-celled animals known as foraminifera have been preserved in the rocks since Permian times (almost 300 million years ago). Ostracods are minute crustaceans related to water fleas; coccoliths are calcareous plates produced by algae and fossilized in huge quantities; and spores and pollen from plants are also important index species.

Conodonts from Devonian rocks in Indiana magnified 12 times. These are the only hard parts of soft-bodied animals that remained a mystery until recently. Their exact relationships are unclear, but the animal probably looked like a young eel.

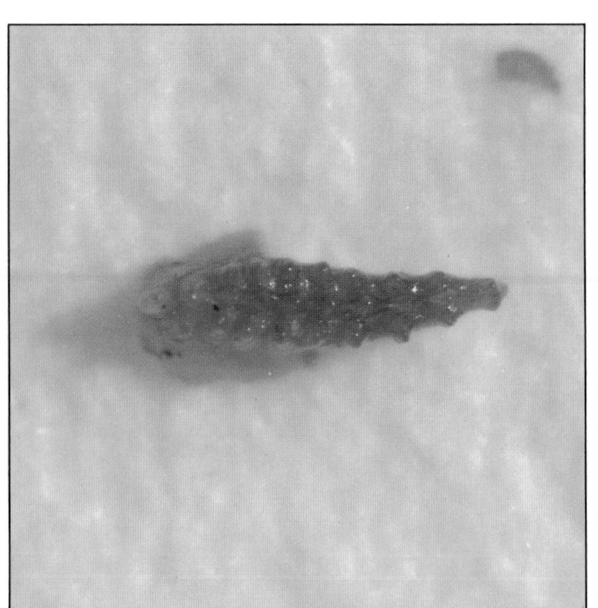

Conodonts are important stratigraphic indicators of Paleozoic and Mesozoic rocks, because they come in many characteristic shapes, and are small enough to be brought up intact in core samples from bore holes. They are extracted from the rock by acids and examined under a microscope.

Left Spores and pollen are very common microfossils because they have been produced in quantity by plants and fungi. They have characteristic patterns on their surfaces, which make them useful as index fossils. This is a spore of *Contagisporites*, a progymnosperm plant, magnified 20 times.

Below Foraminifera fossils from recent seabed sediments, magnified six times. The single-celled animals that made these beautiful shells lived in the sea. When they died, the shells fell to the bottom and formed much of the mud there. This mud, rich in calcium, eventually became chalky rock.

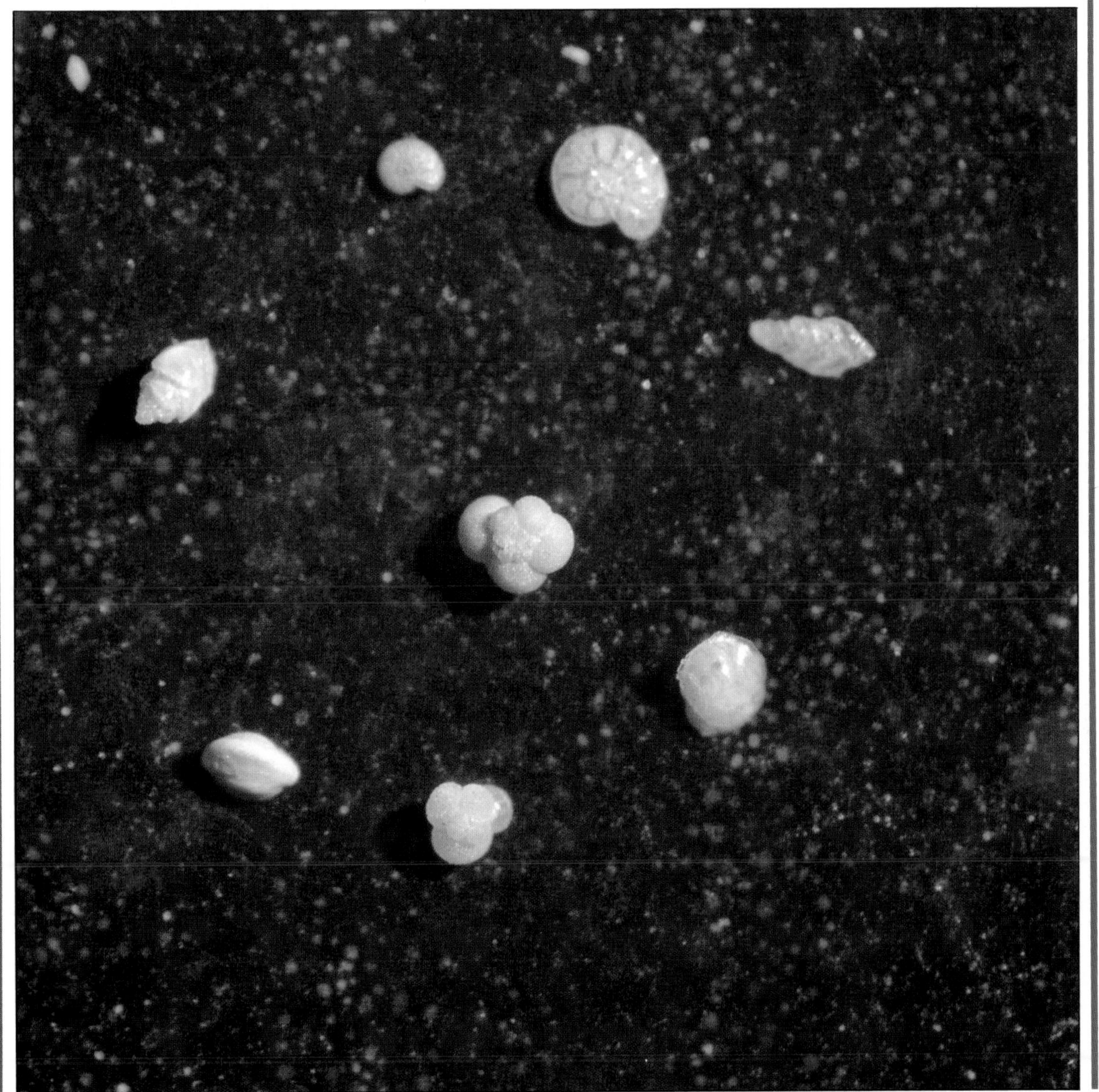

FINDING YOUR OWN LOCALITIES

Collecting and studying fossils, like most activities, makes more sense if it has a theme. A random assortment of unidentified finds from here and there will not suffice for long. It soon becomes apparent that paleontology is a large subject, and it is usually possible only to work knowledgeably in a limited area. By concentrating on one aspect, such as seashells or Cretaceous rocks, you can soon become something of an expert – while picking up broader background knowledge on the way.

What are you looking for?

Consider what aspect of paleontology interests you most, and weigh it against the practicalities.

For many people, it is most convenient to study the fossils of the local region – whatever their age or plant and animal groups. Coupled with geological details, such a study could make a valuable contribution to regional prehistory. In many areas, there are groups you can join, where information can be exchanged. A visit to your local paleontology or geology group is highly recommended.

Regions where sedimentary rocks are not found at the surface – or if they are, they lack fossils – are more frustrating. The amateur paleontologist may decide to make a very thorough search of the area – only to end up confirming there are indeed no fossil beds. On the other hand, careful searches of gravel pits or quarries may reveal new finds.

You may be interested in a particular type of organism, and how it has changed with time. Studies of this nature throw light on the evolution of past life. It may be possible to work in the local area: around the Calvert Cliffs in Maryland, for example, where ancient near-shore sea life once lived in great abundance. But this aspect of paleontology is more likely to involve travel, providing vacation destinations for many years.

Time, money and space

Your choice of study area will also be influenced by the amount of time you can spare, your financial situation and your other commitments. In addition, you should consider where you will store your rapidly growing collection. If space is limited, study small fossils! It does not take many large dinosaur bones to fill an ordinary house, quite apart from the problems of excavating and transporting them. It is of course possible to donate finds – whether large or small – to your local or a national museum.

Planning, and organized trips

Large or small, near or far, your trip will be more productive if it is carefully planned. A useful first move is to visit museums in your area. Here you will see examples of fossils found locally, probably with details of where and when they were recovered, rock types and so on. This information can give you a good start. In regions where fossils are common, fossil stores can provide similar information. The local book store or library should have books on the geology and paleontology of the area. The local museum may house the records of local societies.

A supervised trip to a recognized fossil-rich site can be a great help to you as a beginner. For example, the Jurassic rocks of southern England's Dorset coast, the Cretaceous deposits of the central United States, and the Devonian sites in Canada, are all helpful and encouraging. Training the eye to pick out tell-tale signs – the right sort of rock fragments or surfaces, slight color variations or breaks in the grain of the rock, a certain shape or pattern – can only be done by experience in the field. Ideally, go with knowledgeable guides and learn from them, before attempting to search by yourself. On the whole, fossils are rare and difficult to find. It is less frustrating if your first hunts are successful.

Left Brachiopods are common fossils in marine sedimentary rocks up to 570 million years old. Here, the delicate, internal skeleton of *Spirifer* has been preserved. Brachiopods are numerous and make rewarding subjects for specialization.
Right Waste material from coal mines often contains fossils of plants which formed the coal – ideal subjects if you live near a coalfield. This plant is a Carboniferous horsetail, *Asterophyllites*.

LOOKING AT THE RIGHT ROCKS

Recognizing the rocks where fossils are likely to be concealed is fundamental to paleontology. As part of your preparation, make yourself familiar with the various rock types, using books and the collections of museums, geological groups and paleontological societies.

Most sedimentary rocks will yield fossils of a sort, eventually, but the abundance varies tremendously. On rare occasions, fossils have been found in metamorphic rocks such as slate, which is metamorphosed shale.

The element of chance

Since the beginning of life, there have been chance mud slides, stagnant lakes, upwellings of tar and flash floods. In each case, the conditions for fossilization were different. The finding of a fossil involves the same element of chance that was involved in its creation. Remember, too, that the age of the rocks will affect the state of preservation. The older the rock, the more likely it is to have undergone geological alteration, and the less likely it will be to contain good fossils.

Likely candidates

The origin of sedimentary deposits suggests whether they might contain fossils. Breccia and sandstone deposited on land rarely do, although the Cretaceous sandstones of the Gobi Desert contain dinosaurs like *Tarbosaurus* and *Saichania*.

Sediments deposited underwater are more likely candidates. River sandstone often shows the pattern of ripples, but fast-flowing water only rarely allows rapid burial. Fossils of terrestrial or freshwater origin are very patchy, reflecting the isolation of their habitats. In Alberta, Canada, there are sandstone cliffs which were deposited beneath a river. From these cliffs, over 500 dinosaur skeletons of 50 species have been found. They probably died at a favorite water hole or river crossing.

Deposits under stagnant water, such as bog iron ore, often bear fossils. Shale, derived from the mud deposited by rivers, is another good source of fossils. These very fine sediments settled as the water currents slowed, and often contain organic material and fossils.

Sediments from the bottom of the sea are even better candidates. Limestones usually contain fossils ranging in size from microscopic coccoliths to huge sea creatures. Shells of animals such as arthropods are common, and some of the best fossils have been found in limestone. The beds near the Fitzroy River in Bugle Gap, Kimberley, northwestern Australia, contain numerous well-preserved, three-dimensional fossils of Devonian fish from 370 million years ago. Arguably the world's most famous fossil, the bird *Archaeopteryx*, was first discovered in the fine "lithographic" limestone of Solhofen, Bavaria, in Germany, in 1861.

Chance encounters

Turbidite rocks, derived from catastrophic movements of turbidity currents at steep undersea cliffs, rarely contain fossils – but when they do, they are often spectacular. Some 400 million years ago, on the seabed which now forms the Canadian Rockies, a chance mudslide in a small area trapped and preserved myriad soft-bodied creatures. These mountains had long been known to contain trilobites, but until Charles Walcott, a weary fossil hunter on his way home from a season's collecting, chanced upon fossils of the soft-bodied animals that shared the trilobites' sea, the potential of this

Silts and sandstones on the Makran coast, southern Iran, eroded by the wind to form cliff faces. Fossil hunters develop an eye for a good site, where likely sedimentary rocks are exposed. Any fossils which have eroded out will be found in the loose sand at the bottom of the cliff wall.

area had not been realized. The Burgess Shales are now world-famous. The moral: even well-searched fossil beds often still contain undiscovered treasure.

Coal, although in effect a fossilized forest, does not itself reveal good fossils. But it often contains lumps of shale, within which may be the beautiful patterns of plants and insects. These shale lumps are useless to the mining company and are often piled onto dumps nearby. Peat, too, may contain recent remains of plants and animals.

The finer the grain of the sediment that covers an organism, the finer the detail preserved after fossilization. The limestone of Solnhofen, in the Bavarian region of southern Germany, is exceedingly fine. It is due to the unique nature of this rock that the feathers and other details of the earliest known bird, *Archaeopteryx*, from 160 million years ago, were preserved. Fossils of birds are rare, as the bones and feathers are so fragile, and there is still a gap of many millions of years before the next bird-bearing rocks.

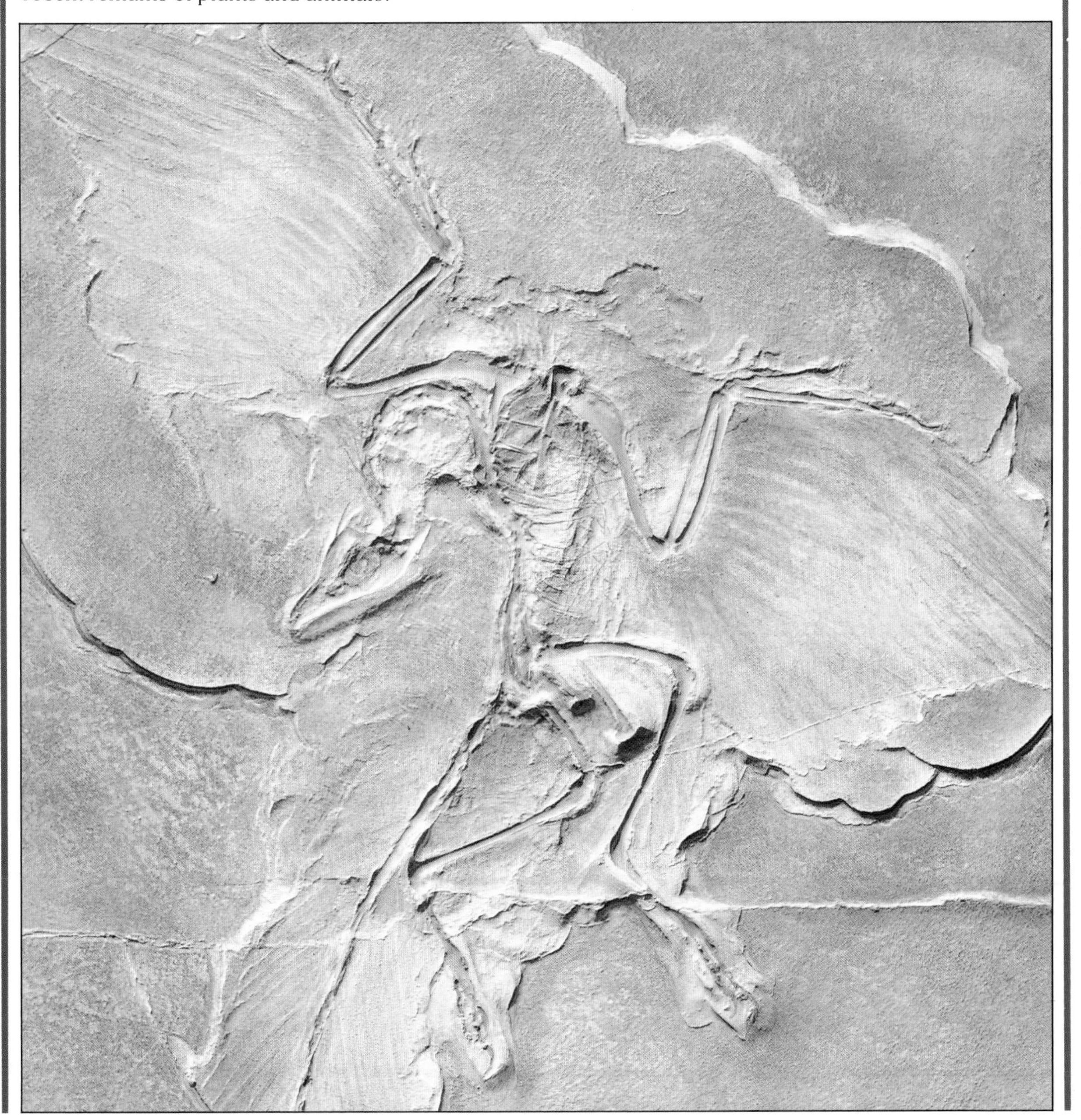

GEOLOGICAL MAPS

Having decided which rocks you wish to search, either for their age or for the types of fossils they harbor, the next stage is to locate the right site with the relevant rocks. Geological maps are invaluable here. They can be obtained from the U.S. Geological Survey or examined at a local museum. The organization that publishes the maps may also produce books or pamphlets covering the geology of a local area.

Poring over a geological map should show you where suitable outcrops occur. For most practical purposes, sites where sedimentary rocks are exposed at the surface are the places for fossils.

The first geological maps were made by Cuvier and Brongniart of Paris, and by William Smith (page 24) in England. They rely on three "laws." First is the Law of Superposition: younger sediments form on top of older ones. Second is the Law of Original Horizontality: water-laid deposits were originally horizontal and parallel to the underlying surface. Third is the Law of Original Continuity: water-laid deposits continued laterally in all directions until they reach the edges of the deposition.

These laws must be qualified, since strata are sometimes tilted off their original horizon by geological movements, or they may be eroded or split at some point.

Using the maps

When using a geological map, remember that the more colorful it is, the greater the variety of rocks. Select the appropriate scale for the sorts of rocks you want. Rock formations are often classified on maps according to geological periods, while the inclination of rock surfaces may also be shown. Landscape features such as dips, strikes, anticlines and synclines, may all be marked. Read an introductory geology text to guide you.

Use the contour lines to look for eroded hillsides. Where the contours are close together and correspond with bands of exposed rocks, there will be a likely site for fossil hunting. The whole map is covered by a reference grid, so you can pinpoint sites exactly and transfer the map reference to a more familiar road map.

If you intend to investigate a site which has been yielding fossils for many years, remember that most of the best remains will already have been

British Geological Survey Map No 161, showing an area of Norwich, England, at a scale of 1:50,000.

Green – Cretaceous
Red – Norwich Crag, Lower Pleistocene
Yellow, blue, pink – Pleistocene glacial
Orange, brown, yellow – recent river deposits

removed. There may also be plenty of attention from other paleontologists searching for recently exposed specimens. It may help to look carefully at a geological map and try to discover where the same beds may be exposed some distance away, and transfer your work there.

There are new laws emerging in the U.S. governing where you can and cannot collect fossils. The conscientious collector should contact and join the national society for his or her branch of paleontology to keep abreast of changes.

What the maps show

Geological maps are a basic requirement for anyone concerned with rock types. They show the following features.

- The rocks exposed at the surface are indicated by different colors, bounded by black lines that are solid if the information is certain, and dashed or broken if it is an estimate.
- Contour lines to indicate the height of land above sea level.
- The main landmarks such as roads, rivers and railroads.
- Sections of the rock beneath the surface soil, together with diagrams of deeper rocks obtained from boreholes, are also shown on some maps.
- Information about gravitational and magnetic forces is included because geological maps are used in the planning of roads and railroads, and for finding mineral deposits of economic value.

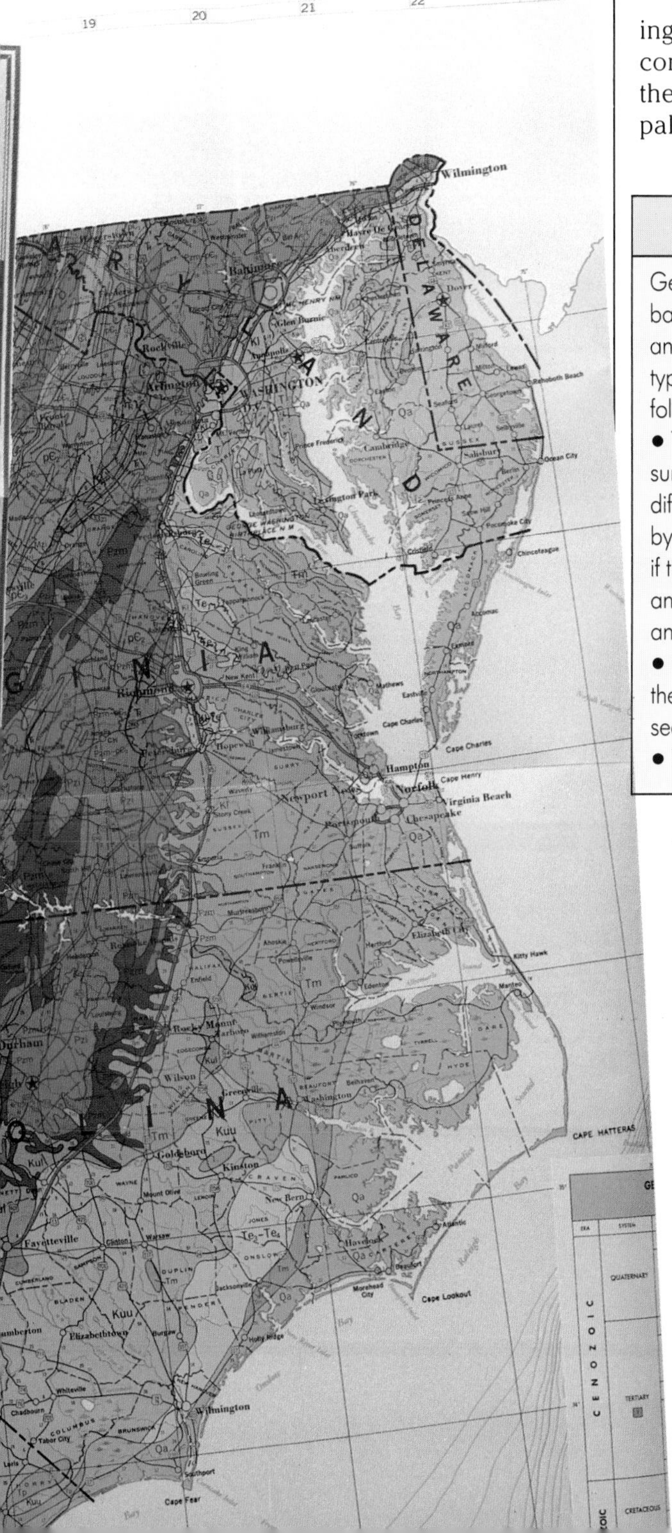

U.S. Geological Highways Map No 4, showing an area of Virginia at a scale of 1 in to 30 miles (2.5 cm to 48 km).

Greeny-brown – Precambrian
Oranges, purples, blues – Paleozoic
Greens – Mesozoic
Yellows – Cenozoic

Today, geological maps are produced with the aid of satellite and aerial photographs. This must be reinforced by practical geological study. The type of rocks and presence of faults and folds are plotted. The structures beneath the surface can be analyzed by drilling boreholes, or by studying the way that electric currents, sound waves and shock waves from explosions pass through the rocks. This study is called seismic (for shock-wave) stratigraphy.

LANDSCAPE FEATURES

Sedimentary rock layers are twisted, bent, thrown into folds and snapped as the Earth's crust moves. The resulting features vary in scale from cracks a few inches wide to whole mountain ranges. Fossils might be exposed where the layers are brought to the surface by these movements. So an understanding of geological folding and faulting is a great help in finding the best area for a dig.

Folds

The different folding processes and their results are described by a long list of geological terms. When the layers of rock are subjected to huge forces, they fold in increasing stages. The flatter regions, or limbs, on each side of a fold make an angle with each other called the interlimb angle. The line that bisects this interlimb angle is the axial plane. The upward "hill" part of the fold is the anticline, and the downward "valley" is the syncline. These terms allow the degree of folding to be described.

Folds may be shallow or deep. They may even fold back on themselves, so that the two limbs are at the same horizontal angle, with one upside down.

Some folds are merely domes or basins. But folds at the surface are of little use to fossil hunters if they are unbroken, since edges of rocks will not be exposed for searching. Only where the strata are broken can hidden fossils be found. Look for folds on the geological map and follow them until they are broken by forces of nature (or excavation). This is a likely place.

Faulting

Faults occur where the folds in the crust break, or fracture, and the two sides move, or displace, in relation to each other. As with folds, a multitude of terms is employed to describe faults. Familiarity with some of the terms helps in the quest for identifying likely fossil sites.

Faults may be vertical, or angled to the horizontal, when the angle is known as the dip. The rocks above the fault are known as a hanging wall, and those below as the foot wall.

The strike is the direction of the fault on the compass, such as a fault line that runs north-northeast. If the movement is horizontal along the strike of the fault, it is known as a strike-slip fault. Where movement is roughly vertical, it is a slip fault. Sometimes two fault lines run approximately parallel to each other. If the central block has moved up, it is known as a horst; a block that has slipped down is called a graben.

An ideal fossil-hunting site: a sandstone cliff face with weathered blocks at the base. Continued erosion should mean a ready supply of fresh fossils at the base of the cliff.

Faults are more productive than folds for fossil hunting. Find them on maps and look for the steeply sloped sections where they are not covered by soil. Often the patterned strata are beautifully exposed, and fossils fall out as the rock erodes.

Exposure by erosion

Having found a suitable bedrock on the map, next look for places where it is likely to be broken and exposed. The weathering action of wind and rain cuts into the rocks, eroding the softest types most quickly, and exposing the layers of sediments.

Water is the main eroding agent. It percolates into rock, dissolving out soluble salts and weakening its structural framework. It fills tiny cracks in the rocks and, on freezing, expands and opens them up, often breaking off lumps. Running water picks up the fragments, and as they bounce along,

Folding and faulting

When the rock layers at the Earth's surface respond to the pressures produced by the shifting material below, they gradually buckle and fold, break up and move against each other. These movements sometimes make it difficult for geologists to follow the rock layers from one area to the next. But they help the paleontologist in his or her quest, because the process forces the fossil-bearing rocks up to the surface.

Folding

A fold is any bend in rock layers. When the fold is downward or valley-shaped, it is called a syncline; when it is upward or roof-shaped, it is an anticline. Sometimes buckling is so severe that the layers fold over on themselves and the strata in the center become completely inverted.

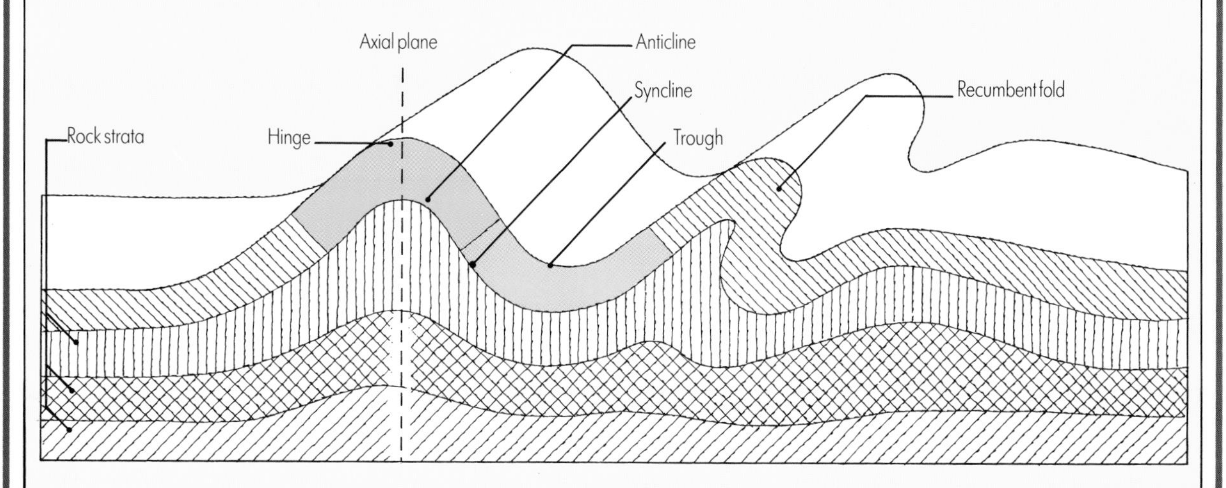

Faulting

Faults occur where the rock has broken and each side of the break has slipped in relation to the other. Different types of fault have different names. When the rock below the fault appears to have moved upward, the fault is known as a normal fault. Relative motion sideways gives a strike-slip fault, the strike being the direction of the fault line. Where there are two parallel faults and the central block of rock has moved upward, the fault is called a horst. The rock below the fault is called the footwall, and that above is the hanging wall. If the central rock between two faults has sunk in relation to the surroundings, the fault is called a graben or rift valley.

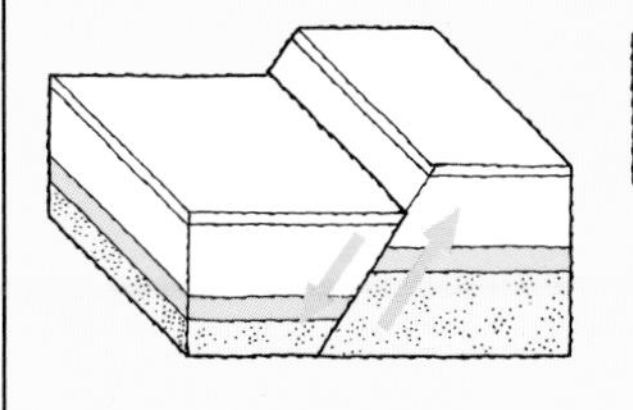

Normal (dip-slip) fault

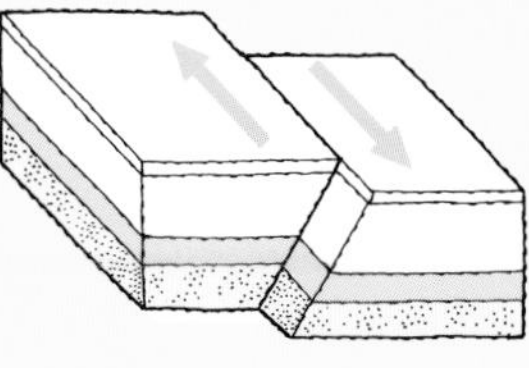

Strike-slip fault

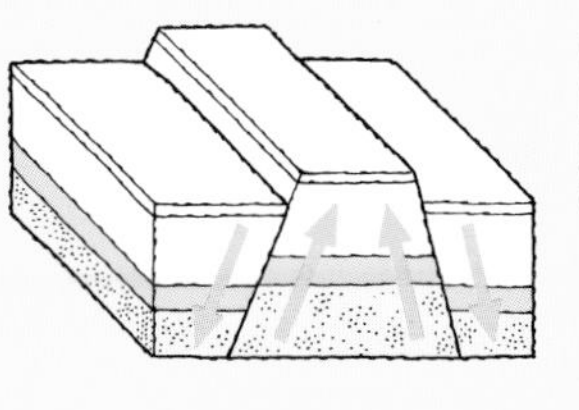

Horst

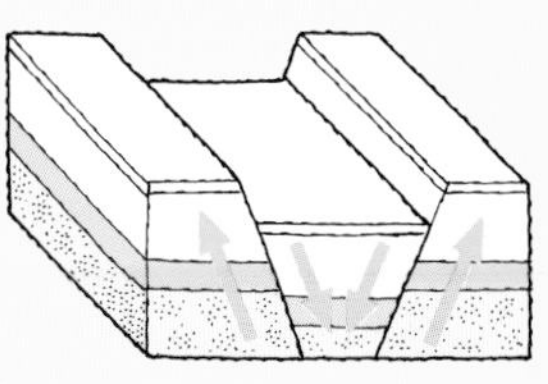

Graben or rift valley

they wear other fragments away from the bed of the stream. Rock beds are often exposed by the action of water, especially fast-moving or large bodies of water. So look for places where streams and rivers cut through and down into the strata. They will often expose bedding planes. In places where ice has traveled across the land in the form of glaciers, erosion takes place because of fragments of rock carried in the ice.

The sea continually eats into the bases of cliffs, throwing water, sand, and pebbles with great force. The rate of wear depends on the hardness of the rock, but since most sedimentary rocks are relatively soft, they wear away quickly and sometimes dramatically. If the cliffs are rich in fossils, new finds are very likely after storms and high seas. So look for places where sedimentary rocks meet the shore of the sea or a lake.

Wind also picks up fragments of rock and hurls them around like a giant sandblaster. Soft rocks, especially high on hills, are worn away by this action. So look for places where rock beds rise at the top of a hill and a weathered outcrop may occur. Desert regions often have such areas of exposed rock.

Human erosion

Our own activities also produce areas where the rock is broken, cut away and exposed. In places where rocks are of commercial value, we dig huge holes with giant machines or drill long tunnels and chambers underground. Peat, coal, metal ores, gemstones, chalk, and china clay; gravel, sand, lime and rock used for building; even fertilizers, are all dug from quarries or mines.

If rocks get in our way, we remove them.

Left The manner in which likely sedimentary rocks are exposed is irrelevant; man-made exposures are as good as nature's. In this chalk quarry in Lincolnshire, England, fossiliferous limestone has been opened to view. Fossils may be found in the chalk faces or in the quarry's waste material.

Below Abandoned railroads which pass through interesting rock beds are also good sites to survey.

Bottom Slate is quarried from the Welsh mountains. Though it is a metamorphic rock, its fine grain and bedding planes sometimes reveal beautiful fossils.

Roads, railroads and canals carve their way across the landscape. These are all possible places for fossils, as are any large scale excavations near the rock beds.

Leisurely surveying

Most of us appreciate a beautiful view across the landscape. But look also with the eyes of a paleontologist. Scan the countryside for likely sites when you are out driving or hiking. Again, look for places where rocks have been exposed by humans or nature. This is another skill which needs a "trained eye," and advice from an experienced colleague is invaluable. If possible, identify and photograph the rock at the site. Or, with permission, take a few samples home and check them in geology books and against specimens, to verify whether the site might contain fossils.

CLOTHING AND EQUIPMENT

Photographs often show professional paleontologists smiling in the sunshine, wearing only T-shirts, shorts, sandals and sunglasses, as they chip away at a valuable find. But this does not convey the whole truth. It does rain at excavation sites, and it does get cold. So be prepared.

For a neighborhood hike on a weekend afternoon, you probably need only a coat and sturdy boots, light refreshments, and of course the practical fossil hunter's best friend – the paleontologist's backpack (page 52). A weekend expedition to a distant site is usually more complicated. You will need transportation. Somewhere to stay might involve lodging with friends, a campsite, or a hotel.

The "vacation expedition"

For a more adventurous expedition, you might link your fossil finding with your vacation – even a week or several abroad. You can sign up for a professional paleontological expedition through an organization called Earth Watch. This involves all the usual preparations such as passports, visas, permission, currency, vaccinations where necessary and insurance for you, your vehicle, equipment and so on. Select a suitable season when the weather is likely to be neither too hot nor too cold – cracking rocks in the glaring midday sun does not suit everyone.

Reservations for transportation and accommodation can be made through your local travel agent. Plans for food and drink normally involve local stores, cafés and restaurants. However, if your dig is some distance from the nearest stores, you should stock up with adequate rations.

Clothing

Clothes and footwear should suit your trip's weather and conditions. Most of the guidelines for hikers and mountaineers also apply to the paleontologist.

Warmth is vital. You can always remove layers of clothing if the sun comes out or if you work up a sweat while you are working. But when it gets cold, you cannot put on sweaters and coats you do not have. The chief rule is: clothing must be warm, windproof and waterproof. Several thin layers are better than one heavy garment. A warm hat, thick socks and gloves (see below) are needed for cold-weather trips. A folded coat can double as a kneepad if you kneel for long periods.

The well-equipped fossil hunter, thorougly absorbed in his work. Protective clothing makes a collecting trip safer, and suitable equipment eases the extraction of fossils.

In hot conditions, cool, light-colored garments are ideal. Avoid artificial fibers, such as nylon, which prevent perspiration from evaporating and do not "breathe." Wear a sunhat and keep your shoulders and back covered, at least for the first few days; it is all too easy to get sunburned without realizing it as you crouch over some interesting find. Sunglasses help to prevent troublesome glare from light-colored rocks.

On your feet

Footwear should be adequate for your exploration. Walking or climbing boots may be necessary, but even if you do not need these, you should protect your feet from the tools of the trade. Sturdy boots or shoes minimize injury from hammers, chisels, spades and falling rocks. Leather hiking or work boots are best in wet places, although they may be uncomfortable to wear for long periods.

Safety

Gloves are usually necessary to protect your hands when digging. Hard hats or safety helmets are a must for work involving overhead rock faces, cliffs, quarries or cuttings. A single small rock can cause serious injury when it falls some distance.

If you need to climb cliffs or mountains to find your fossils, it is essential to have proper training in climbing. Wear suitable gear, and take the proper equipment. It is safer to go with an experienced climber (see page 54).

Equipment for finding, extracting and transporting the fossils themselves is discussed elsewhere in the book (see page 52).

Preparation for your expedition depends on where you intend to find fossils. Previous training and suitable equipment and clothing are necessary for trips that involve potentially dangerous places, such as caves or cliff faces. Fossil hunting, however, combines well with such sports as caving or rock climbing.

Fossil collecting in remote areas may involve more specialized preparations and equipment. This temporary bridge in Tanzania has been erected solely for the fossil-hunting expedition to the area. Such vehicles are necessary where roads are not paved.

GETTING PERMISSION

Most of the land surface of our planet is owned: by a government, organization or individual. You should *always* find out who owns the area where you wish to search for fossils and ask for permission. If it is public land, you must secure permission from the Bureau of Land Management. Not only is it courteous, it could save you from embarrassment later; it could even stop you from going to prison. Sometimes the "Trespassing forbidden" sign has little foundation in law, for the hiker. But as soon as you start chipping at rocks, you could be damaging another person's property, and this can have more serious consequences.

The legal position regarding public access differs from one area to another. You must find out the position in the place where you wish to search. In large, sparsely populated regions, many landowners are not too concerned about members of the public wandering the vast open spaces. Australia, New Zealand, Canada, Scandinavia, and parts of the United States, Africa, Asia and South America have huge unpopulated areas where fossiling is unlikely to interfere with anyone. In smaller countries there are far fewer places where landowners have no interest in what goes on.

Fossil hunting in regions where precious stones, metal ores, or other minerals are being mined could be interpreted as prospecting, and therefore liable to punishment. This applies especially to parts of the United States, South Africa, the Soviet Union and Australia.

Finding the landowners

If the land is the property of the government or an organization, there may or may not be signs at the boundaries identifying ownership, and telling you who is permitted to use the land and for what purpose. If there are no signs, ask at a nearby house, store, post office or bar. If this fails, make a trip to the nearest local records office or museum. When you find the name and address of the landowner, a polite letter or telephone call stating your aims and reasons may be all that is necessary. In general, landowners – if approached properly – do not object to fossil hunting.

Even when there is access to "public" areas such as footpaths, mountains or beaches, your rights may not include excavating and removing lumps of rock. The municipality should have copies of local laws covering such activities.

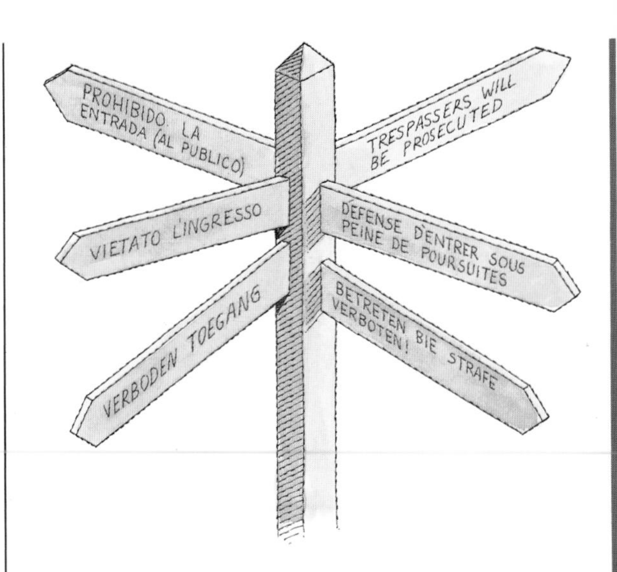

The laws against trespassing differ from one country to another. But the chances are that a landowner will be more amenable if a courteous request is made before a fossil hunter starts removing property.

Quarries and construction sites

At quarries and construction sites, and on farms and mining sites, there is usually a manager to contact for permission. Often, large machines are operating, and it is doubtful if the manager will agree to hold up work or take responsibility for your safety. But he or she may not object to your return when the machines are not working – provided that you ask. You may have to agree to wear a hard hat and obey safety regulations while on the site.

Quarry managers may be used to fossil hunters searching the area. They might even suggest likely places to look. If you are lucky enough to find anything that may be of scientific importance, tell the quarry manager and get in touch with the local museum. Quarries and mines have sometimes stopped work to allow excavation of important fossils. Housing projects in many parts of the U.S. are required by law to allow fossil collection by professional paleontologists.

Forbidden areas

Some places are inaccessible because they are used by the military – they may be both secret and dangerous. Nature reserves and conservation parks have only limited access, in order to preserve

Top It may not be possible to investigate a quarry while heavy equipment and workers are on site. Ask the manager for a suitable time and place for fossil hunting.

Above If your quarry-searching is successful and you have the goodwill of the site manager, work might stop in that spot while you ask experts to examine the site.

the animals, plants and rocks that are found there.

Some of the best fossil sites have been made national monuments such as La Brea Tar Pits in Los Angeles or Dinosaur National Monument in Colorado and Utah. At such places, only professional paleontologists from eminent museums or universities are allowed to carry out excavations. If you wish to see such a place, you may be able to go along as a sightseer or volunteer.

Remember...

- Find out who owns the site.
- Get permission from the landowner, manager, or organization, such as the National Parks authority.
- When going abroad, write to the embassy of the country you wish to visit for an international permit or advice on who to contact.
- Be aware of landowners' rights and responsibilities, and respect their wishes.
- Negotiation and diplomacy are better than trespass and liability.

FINDING YOUR FOSSIL

The tools and equipment used for fossil hunting are similar to those used by geologists for rock sampling. This equipment will help you find the fossils, record their positions, carefully extract them, and then label, wrap and protect them.

Most fossils are solid rock; they are not light, and the farther you carry them, the heavier they seem to get. The tools needed for excavation are also quite heavy. For big finds, you may need help, and perhaps even a wheelbarrow. But beware of filling your backpack too full. Take only the tools that will be absolutely necessary. What you choose will depend on the type of dig, and how far away it is. If you can drive and park close to the site, the heavier items can be left in the vehicle.

The paleontologist's backpack

A backpack is invariably more suitable than a shoulder or hand-held bag. It carries more, more securely, and in a more naturally balanced way that is less tiring to the body. In addition, a backpack leaves both hands free in case you fall. And if you are in a group, you can take turns to share the load.

The backpack itself should be fairly roomy and comfortable to wear, preferably with padded straps that spread the weight evenly. Stores that specialize in climbing and camping usually have a selection of suitable backpacks. Modern nylon ones are exceptionally strong and do not rot in damp conditions.

To locate the site

The items listed here will help you to locate the fossil site, and will help to get you home again.

- A good, large-scale standard map (minimum scale 1:50,000), plus the relevant geological map, folded to show the area in question and protected inside a transparent map-carrier.
- A magnetic compass.
- Tidal charts for fossil hunting under beach cliffs.
- Letter of permission from the landowner.
- Pocket guides to rocks and fossils.

Checklist of tools and equipment

You will not need all these tools every time you visit a dig. Take only what is necessary for the type of work at hand and the stage of excavation.

- A small lightweight spade for digging out loose material.
- A geological hammer, with one square end and a pick end or chisel at the other; weight around 1-2 lbs. (0.5-1 kg). It should have a long loop of cord with a knot. The loop is threaded through a hole in the end of the handle, so that it can be wound double around your chest and shoulder for carrying, and looped singly while working, to prevent dropping and losing the hammer.
- A lump or club hammer. This may be useful for hitting chisels, but this is usually too heavy to carry long distances. Other possibilities are a small chipping hammer, a large rock hammer or a geological hammer.
- A rock saw and stonemason's chisels, of assorted blade widths from ¼-2 in. (0.5-5 cm), for working around a fossil.
- A brush, such as an old toothbrush or shaving brush, for removing loose rock.
- A small trowel or an old knife to scrape away soft rock.
- A strainer for removing small fossils from friable sand or clay, or for washing a specimen in a stream or lake.

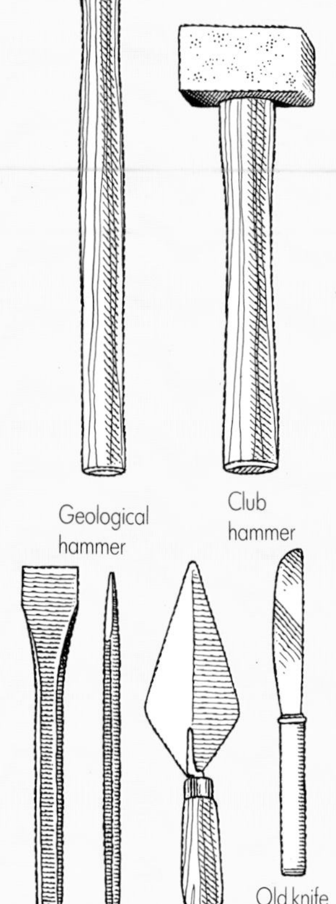

Paintbrush

Shaving brush

Sieve

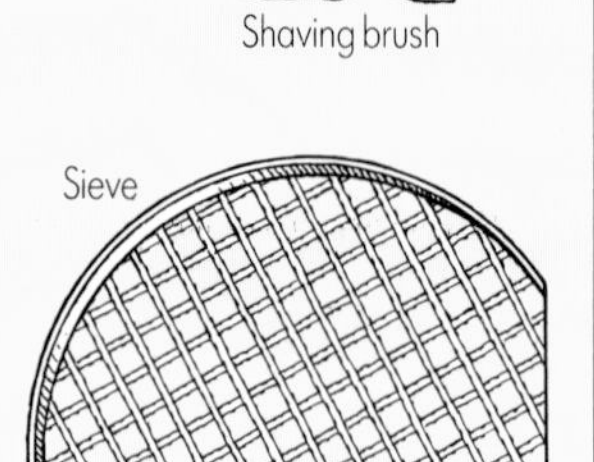

A party of fossil hunters winds its way through the Dinosaur Provincial Park, in the Badlands of Canada. Outings to such famous sites are carefully organized by guides. Collecting may not be allowed, but suitable clothing is still necessary and equipment such as a hand lens, map and field guide help the amateur get the most from the expedition.

TAILORING YOUR TRIP TO YOU

The sites where you dig and the types of fossil you aim for should depend on your abilities. Set your own limits and stay within them, to avoid disappointment. Are you fit enough to walk long distances and spend many hours hammering at rocks? Do you have the time to journey to and from the site, leaving a good period for searching when you get there? If you take young children, will they quickly become bored and need some form of entertainment?

If your physical capabilities or mobility are limited, for whatever reason, then choose the types of sites and rocks that are likely to yield fossils easily, without much effort. Beaches or quarries under shale and fossiliferous limestone cliffs are good sites. Looking for surface flints or ironstone nodules is gentle exercise. Searching the surface of mudstone rocks for tracks, trails, burrows and tunnels can be rewarding.

If you are looking for something more strenuous, fossil hunting can be an excellent pursuit. Long treks in awkward terrain, hill walking and climbing can all be enjoyed in the pursuit of specimens.

The skills of climbing

Climbing is a skill often needed by advanced paleontologists. But you must learn how to do it properly. The dangers are obvious, but even so, they cannot be over-estimated. So if this style of fossil hunting appeals to you, take a course in basic rock climbing. The correct equipment is also essential. Climbing ropes, harnesses and clips are made of special materials, very strong for their light weight. A piece of old rope from the garage will *not* do; it may break at the most dangerous point. Remember also: do not climb unless it is necessary; do not climb on unsafe cliffs; do not climb alone; and do not attempt climbs beyond your own capabilities.

All this involves time and expense to start with, even before you set eyes on a fossil, but it is money well spent. If you are new to paleontology, it would be advisable to begin at sites which are more easily accessible, to see how you take to it. Then you can progress to digs at more challenging sites.

Homeward bound

Always bear in mind that you will probably want to take your fossils home, or at least to a workroom or laboratory somewhere. This also sets limits on where you search and the types of fossils you attempt to extract. If you are collecting small fossils, do not fill your bag too full. Without damaging the fossil, chip away as much of the rock as possible, to avoid extra weight (page 64).

It is easier to carry a heavy load over a paved road than along a rough track or plowed field. If necessary, leave your find in a safe place and return later with help and transportation.

Searching weathered Triassic rocks in Namibia. Sifting through fine, dusty deposits in the hot sun does not suit everyone. Collecting large amounts of rock fragments presents problems of transportation for the amateur. Analysis of such material will require time and laboratory facilities.

Above These huge blocks of Portlandian rock, near Boulogne, France, are covered with the casts of invertebrates' burrows. Collecting specimens from such rocks involves chipping a suitable-sized piece using a geological hammer, here carried on the belt.
Left The paleontologist's backpack as used in Namibia. Large fossil-bearing fragments of rock are being removed, brushed and wrapped in newspaper before being carried away in the backpack

THE INITIAL SITE SEARCH

As emphasized throughout this book, almost anyone can go fossil hunting – but fossil finding is a skill that depends largely on experience. Outings with an expert will certainly help you to acquire this skill, but practice is still necessary. True, some people are naturally better beginners than others, and they soon pick up the knack of spotting a good specimen. If you are not one of them, you can still improve your chances in various ways.

Keeping your study area small and well defined – whether by the type of site, the age of rocks, or the groups of fossils – usually means that you become an expert more quickly. Remember that whole fossils are very rare, and you are mostly likely to find bits of animals and plants. Teeth are common in rocks bearing vertebrate fossils; pieces of bone survive rather than whole skeletons; creatures like crinoids tend to break into short columns; bark fragments are more likely than whole plants.

Once you have established that a site does, or should, contain fossils, the next stage is to carry out an initial search. This indicates whether it is worth investing the time needed to set up a proper, more organized dig.

Beaches and cliffs

Beachcombing is a pleasant way of looking for fossils. If the beach is adjacent to fossil-bearing cliffs, specimens should be easy to find among the rocks and shingle on the shore. Fossils that weather out and fall to the beach are often hard enough to resist fast erosion by the action of wind, water and sun. These types of fossils are likely to remain intact for some time.

Keep your eye open for something unusually smooth, or geometric, or regularly patterned, that catches the light. It may be the internal surface of an ammonite shell or an echinoderm skeletal plate. The curve of a "pebble" may be too perfect to have been worn by the sea – it could be part of a brachiopod shell. Belemnite "bullets," the internal skeletons of cephalopod-like animals, often occur in large quantities, while the thick spines of certain

The absorbing hobby of beachcombing for fossils has fascinated both young and old for centuries. At suitable sites, this relaxing pastime may provide enough interest for the part-time fossil hunter – and the rest of the family.

Above This beautifully sculpted sea urchin shell, *Phymosoma koenigi*, was found near the Cretaceous chalk cliffs of Dover in Kent, England.
Left Large fossil ammonite shells in the Jurassic rocks on the beach at Lyme Regis, Dorset, England. The sea has worn the rocks so that the shells are seen in section.

sea urchins look very similar. These characteristic shapes, once seen, catch the eye again and again.

Watch also for changes in color, for patches of pigments or outlines that look "biological." Carbonized fossils are black or dark against the lighter background. Calcified fossils, tunnels, bryozoans and graptolites are often light against the darker background rock.

Cliff faces are being eroded all the time. The rate depends on the hardness of the rock. As the matrix (the rock around the fossil) weathers away, pieces of tougher fossils stand out from the surface. Eventually lumps crack off or fall out and drop to the bottom of the cliff. Walking along the cliff base is therefore a good way to find remains. Bad weather, wind and storms increase the rate of erosion, so try to get to the shore during the calm period after a storm. If it is a popular site, make sure you are early.

Climbing rock falls

Climbing on dangerous cliffs may not be for you. But you may wish to discover which layers, or strata, the fossils have come from, in order to find out more about their age and how they were fossilized. To do this, climb carefully up the rock fall, looking for the same types of fossils as you go. Just below the place where you stop finding such specimens, you should find the seam that they are coming from. Never forget that loose rocks can be extremely risky: always bear in mind the safety considerations when climbing.

Fields and farmland

Flint nodules often lie on the surface of the soil. You may happen upon them simply by walking across fields. They are formed from the skeletons of sponges, and in certain areas they contain specimens of whole sponges, together with the organisms that shared their seas: sea urchins and various mollusks. Break them open with a small hammer; you may have to crack several before you discover a fossil, but it may be beautifully preserved when you finally locate it. In some areas these types of remains are so common that they have local names, such as "fairy loaves," "fairy hearts" and "shepherd's crowns" (the name often given to sea urchins).

Quarry talus slopes

At rock faults, quarries, mines and on the seashore, search rock debris for lumps of rock that may contain fossils. Ironstone nodules from mining waste sometimes contain fossils of fish or insects; they can be broken open with a hammer. Chunks of calcium carbonate are also found in coal seams and may preserve the internal structure of the plants that made the coal. Slate sometimes contain fossils. Scan excavation sites for pieces with a slight bulge on the flat surface. Breaking the slate along its natural bedding plane may reveal a fossil. Remember to ask permission, and be very careful – talus slopes can slip.

Surface features on the rock

In hard rocks, where erosion is very slow, you will have to search the surface of the rock for tell-tale signs of fossils. In material such as fossiliferous sandstone, they are often easy to find, and almost

any lump could contain numerous fossils. In less productive rocks, resist the temptation to hammer indiscriminately. You will probably end up with a pile of rock fragments but no fossils. Instead, examine the rock surface carefully for the bedding planes or lines that indicate how the rock was laid down. Bedding planes demark shifts in types of sedimentary rock, and fossils are often found at these bedding planes. The material will usually break easily along these lines, and fossils are more likely to lie along them.

Softer rock, such as chalk, can be scraped with a chisel or trowel; it does not have bedding planes as such. Carefully work away at the surface and around any suspected remains until you can lever them out. Soft chalks and clays may only reveal their fossils when separated with a sieve. Take samples home in bags and analyze them carefully under a magnifying glass.

Should I persist?

If you think the rocks should contain fossils, keep trying. Go back again another day, with eyes and arms refreshed, and take another look at anything which seems organic. Fossils often occur in pockets, with plain rock in between. The next pocket may be only a few inches away, and it may contain a startling new find.

Stegosaur bones waiting to be freed at the National Dinosaur Monument, Jensen, Utah. Only professional paleontologists excavate bones from such sites, but the amateur will find a visit both fascinating and instructive.

SURVEYING AND RECORDING

Once you have decided that a particular site is worthy of sustained attention, the next priority is to plan a systematic search and survey. Cover the ground or hillside by crossing from one end to the other, and then return along a line slightly to one side of this; repeat the process. This means you will scan the area thoroughly without missing a part or covering an area twice. Make notes as you go of anything that may be significant: changes in the color of the rock, bedding planes, weathering patterns, landmarks and so on, may all help when you try to place your fossil. Take careful measurements so that the depths of rock strata are known, and the scale of any fossil finds can be seen.

The stratigraphy of the fossils is important and you should certainly take the trouble to record it. The rock stratum in which the fossil lies is the best clue to its age. It may be very narrow, so a careful and close examination is usually necessary. Other fossils in the same or neighboring strata allow you to date your fossil more accurately.

The importance of sketch maps

Make a sketch of the site as seen vertically, showing the different sorts of horizons (recognizable time planes) exposed. Note the features of each rock layer: grain size, friability, color, fossil content etc. You should also include careful measurements for scale. Note on the sketch any landmarks which can be used as a reference, to help you find the exact site again, as well as the precise position of your fossil – or where you think it was weathered from.

A camera can be very useful, to record accurately the position of features, types of rock and landmarks. The advantage of a camera is that numerous shots, from long-distance to close-up, can quickly be taken. Again, remember to record in your notebook the frame number of each shot, and what it is supposed to show. It is very difficult to remember such details when the film is returned after it has been developed.

Some differences between rock layers, for example slight variations in texture or color, are too subtle to be captured on film. In addition, shadows can obscure important details, which your eyes compensate for and which you therefore notice. Moreover, you will not have any idea whether faint features will appear on the photograph until it is processed. For many such reasons, the camera is no substitute for pencil and paper. Photographs *must* be accompanied by notes and sketches.

You may find it difficult to identify certain rocks on site. If so, take some samples (recording their position) and use books or specimens at home to identify them. If you are still unsure, your local museum should be able to help.

Why positional records are vital

The position of a large fossil or a collection of fossils, for example from a skeleton, must be recorded. The position in which the animal is lying yields information about its shape, and how it lived and died. The positions of the bones relative to each other form the basis for reconstruction of

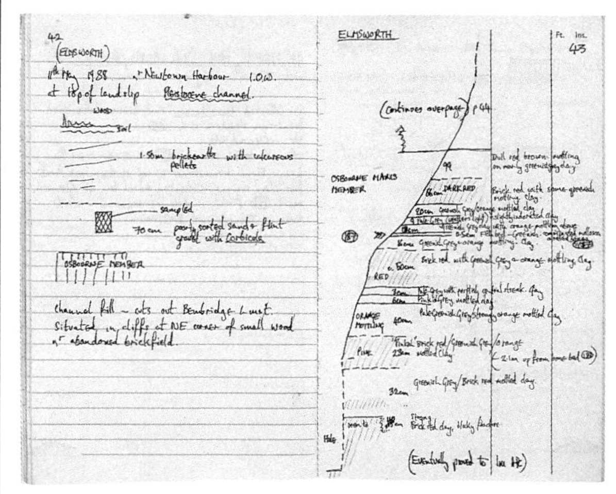

Sample pages from a paleontologist's notebook. The contents include the date and location of the find, a sketch map of the site and descriptions of the various types of strata.

To record the find

To record the position of the fossil you will need the following items.

- A measuring tape or folding ruler, to record the position of the fossil in relation to the rock layers (this helps to date the specimen).
- String, to make a survey grid (page 63).
- A camera, to make an instant record of the fossil's general position.
- A magnifying glass, for examining small finds.
- Notebook and pencil, for recording data and sketching close-up details.

the skeleton (page 82). The positions of other bones, shells, scales, or leaves may indicate where the animal's stomach was, and what it ate, or the creatures and plants that were its contemporaries. Its cause of death, and whether scavengers feasted on its carcass, may also be revealed. Experts are sometimes able to propose a death-and-preservation scenario, such as falling down an underwater cliff into stagnant deep-sea ooze. The study of these processes – how a fossil comes to be buried – is known as taphonomy. Once the fossil is excavated, this type of detail is lost unless it has been well recorded.

Grid methods

The standard way of recording the positions of finds is by the use of a grid. It is a technique widely used in archaeology and geology.

The principle is simple. The area containing the partially exposed fossil is criss-crossed with strings, tied to pegs around the edges, and set out to create one-yard (meter) squares. A wooden frame one yard (meter) square, divided into 100 squares each measuring four inches (10 centimeters) is then moved along the string grid, square by square. The position of each part of the fossil, or an individual fossil, is then recorded using a map-like grid reference. The contents of each square can also be drawn onto graph paper. The string can be removed for excavation, but its position on the pegs should be marked, and the pegs themselves must not be disturbed.

If the fossil's size warrants it, the whole process is repeated after excavation at a lower level – for example, four inches (10 centimeters) down. This necessitates a vertical scale to one side of the study area, which can be projected onto the area with a spirit level or surveyor's equipment. For a really big discovery, a three-dimensional record is built up at several levels as the dig deepens.

This process is a skilled job, and it takes time. If you do find a large fossil, such as a dinosaur bone or giant ammonite, it is unlikely you will be able to excavate it alone. The scientific importance of the find may demand that you inform the experts, so that all the necessary information from the site can be recorded before the remains are recovered. So notify the local natural history museum. You may be asked to show them where the fossil is, and you may be involved in the excavation.

Be safe, not sorry

In the field, safety is a prime consideration. Take every possible precaution to prevent accidents while you are fossil hunting, and be prepared to cope if they do occur.

- Make sure someone knows exactly where you are going on your trip, and how long you are likely to be. They can call the police if you do not return on time.
- Bear in mind the dangers of a particular site, and take notice of warning signs. Cuts along working railroads or highways are out of bounds; the dangers are obvious. Cliffs may fall at any time with no warning, particularly after rain. Abandoned quarries and mines are very dangerous places; old workings and equipment or rock piles may collapse at the slightest disturbance.
- Any water is a potential hazard. Water that fills extraction holes may be very deep, and the hole itself steep-sided. At the shore, the land may shelve steeply. There could be strong undercurrents, or the tide might cut off an escape route.
- Beware of other dangers, such as cliff-falls, mudslides, quicksand, and local poisonous animals and plants.
- Going with someone else is both safer and more fun.
- Brightly colored clothes are more easily spotted by searchers if you become lost.
- Carry a small first-aid kit if you are far from home. The kit, in a light but airtight box, should contain individual adhesive dressings, sterile eye pad, triangular bandage, sterile coverings of various sizes, and safety pins. Small scissors and tweezers may also be useful. A trained first-aider is essential for a bigger expedition.
- Always carry an adequate supply of drinking water.
- Carry some identification, with a telephone number to contact in case of an accident. Keep some change handy for a public telephone, and be familiar with the number for the emergency services.

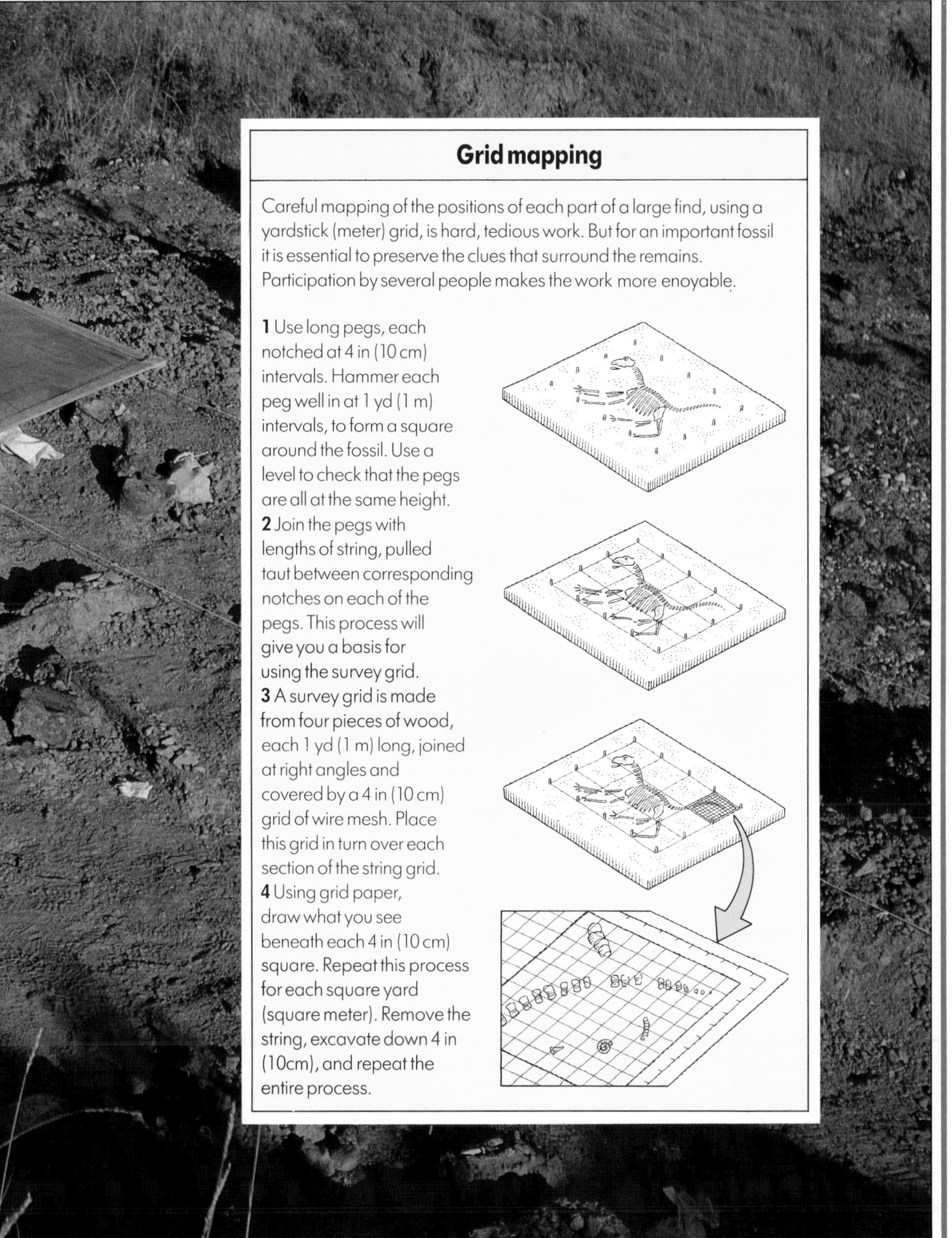

Grid mapping

Careful mapping of the positions of each part of a large find, using a yardstick (meter) grid, is hard, tedious work. But for an important fossil it is essential to preserve the clues that surround the remains. Participation by several people makes the work more enoyable.

1 Use long pegs, each notched at 4 in (10 cm) intervals. Hammer each peg well in at 1 yd (1 m) intervals, to form a square around the fossil. Use a level to check that the pegs are all at the same height.
2 Join the pegs with lengths of string, pulled taut between corresponding notches on each of the pegs. This process will give you a basis for using the survey grid.
3 A survey grid is made from four pieces of wood, each 1 yd (1 m) long, joined at right angles and covered by a 4 in (10 cm) grid of wire mesh. Place this grid in turn over each section of the string grid.
4 Using grid paper, draw what you see beneath each 4 in (10 cm) square. Repeat this process for each square yard (square meter). Remove the string, excavate down 4 in (10cm), and repeat the entire process.

EXTRACTING YOUR FOSSIL

Sometimes, a hard fossil simply falls from the soft surrounding rock. All it then needs is cleaning with a brush. But many fossils are not so easy to collect. The rock they are embedded in – the matrix – is hard, and the remains require many hours of painstaking chipping, chiseling, scraping and brushing in order to extract them. At every stage, there is the possibility of gouging the fossil's surface or cracking it right through. You may need to clear away vegetation, soil and weathered rock on overgrown sites to expose a clean rock face before you can start. Use a pick-mattock for this work.

Extracting a fossil in sediment

Do not assume that you must extract your find then and there, on site. For smaller fossils, it may be better to work out or free a large chunk of rock that you are certain will contain the entire specimen. You can transport it back to your workroom for the final, careful stages of extraction. This method is preferable to chipping off the matrix quickly in the field.

Your planned method of final extraction and cleaning may affect the way you handle the specimen on site. If you are dealing with a rock that has abundant fossils, such as fossiliferous limestone or chalk, or if you are looking for microfossils, chisel off a specimen lump and take it home to be treated with acid (page 73).

When dealing with familiar types of remains, you can compare the dimensions and curvatures of the exposed portion, and estimate where and how far the fossil extends into the rock. However, for a more unfamiliar organism, try to expose as much as possible of it *in situ*, so that you can assess its size. This minimizes the risk of separating a large chunk of rock, only to find that your prize has been split into two and part of it still lies securely embedded in the matrix.

Hammer and chisel

Learning the skills of using a hammer and chisel requires time and patience. Practice on unimportant chunks of rock at first, in order to learn the safe, comfortable and controllable way to grip a hammer and chisel, and how to direct the hammer blows. Then search the site for a few pieces of the same rock in which your fossil is hiding, or for a less significant part of the face made of such rock, and have a practice session. Gauge the direction of the rock's lines of weakness, and the amount of force needed to split off flakes.

The chisels are for working out chunks of rock or for splitting rocks apart. Plan your strategy for extracting the fossil. It may help to imagine that you are a sculptor, chipping a block of stone to the size and shape that you have formulated in your mind's eye. Indiscriminate hammering may at best do unnecessary damage to the environment; at worst, it could destroy an irreplaceable specimen.

Fossils can often be found inside nodules of flint or ironstone. A sharp blow with a hammer may break the stone and reveal the treasures inside. Begin with a fairly light action, and gradually deliver heavier knocks as required. If you do not hit the stone too hard, it should break around the fossil, rather than through it. If a hammer blow separates the rock around a fossil, remember to keep both halves – the mold and cast sections – in order to obtain maximum details.

Choosing the break point

To break an individual piece of rock at a desired place, first study it carefully to locate the bedding planes and cleavage lines. Support the rock so that it will not fall and damage the fossil. Then hit it sharply with the chisel end of your hammer along

Precautions to take

During the initial stages, do not work too close to the fossil itself with your tools, or a slip could damage it. And be aware of the standard safety measures to avoid hurting yourself.

- Wear shatterproof glasses or goggles; shards of stone have a habit of flying up into the face.
- Wear gloves, too, for heavier work.
- Stop every few minutes, brush away the debris, and re-examine the remains.
- Note any newly uncovered parts of the specimen, and if they are not as you predicted, adapt your plan accordingly.
- Beware also of chiseling into a more important fossil in your haste to free the specimen you noticed first.
- Above all: do not rush. Your fossil has been in its rock for millions of years. A few more hours or days makes little difference. Nature has spent millennia fashioning the specimen; hasty extraction could ruin it in a few seconds.

Above At this stage of the excavation, it is difficult to be sure what the fossil may be. However, you may have some idea of the likely finds where you are searching. These heavy bones are those of a mammal-like reptile from Triassic rocks in Namibia. These animals, called therapsids, were the dominant vertebrates immediately prior to the evolution of the dinosaurs. This one is a dicynodont, a herbivorous animal.

To aid reconstruction later on, rough sketches are made of the arrangement of the fossil. **Right** A sketch taken from the fossil site shown above. **Below right** After much excavation work and restoration, a complete skeleton could be reconstructed, despite the unpromising appearance of the initial find. This reconstruction shows the dicynodont *Kanneymeria*, which grew to about two yards (two meters).

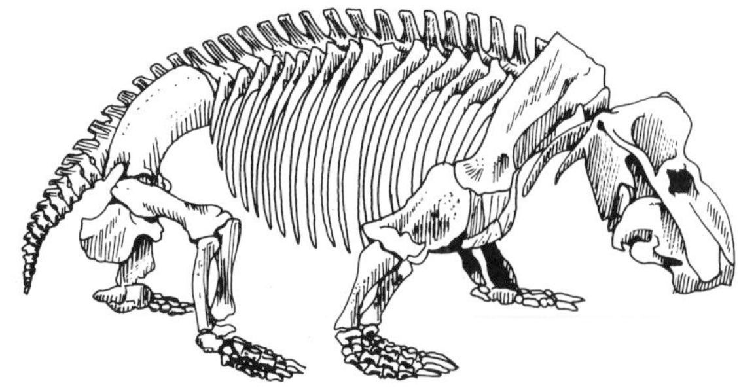

the bedding plane, or with the blunt end if there is no obvious bedding plane. Different types of rocks split in different ways: shales have definite bedding planes, while chalks may have none. Again, practice on disposable nodules before trying the piece that you think may enclose a fossil. If your blow is too gentle, it merely makes a dent; too hard, and it may shatter the piece, and possibly any fossils it contains.

Older rocks may have been subjected to forces which have caused cleavage in a different direction from the original bedding plane. Attempting to extract fossils from such rocks can be very frustrating, even for the expert. Splitting the material along the newer cleavage plane wil! give only a transverse section of the fossil; as this is likely to be very thin, it is easy to miss. So examine the break very carefully for traces of changes in color or texture along the original bedding plane, and try to split the rock along such a line, but away from the possible fossil. Or look for a place where the folding of the rock causes the bedding and cleavage planes to run parallel. A few hammer blows here may be more effective.

Opposite page
Top Fossils are usually found embedded in rock, most of which must be removed for study and display. Even these huge bones of *Antrodemus* (or *Allosaurus*) must be very carefully excavated. Here a preparator at the National Dinosaur Park, Utah, picks away rock with a hammer and small chisel.
Below left Rocks with an obvious bedding plane, such as shale, contain fossils along these fracture lines. The skill of driving the chisel into the correct bed comes with practice. Look carefully at the edge of the rock and aim for any irregularity in color or texture.
Below right As soon as a delicate fossil is exposed, it should be painted with hardener and allowed to dry. Chipping away more rock could easily damage the fossil. Using hardener takes time but prevents a disaster. Here the fossil is in loose rock and would disintegrate if not hardened before removal.

Right This piece of rock contains several bone fragments, as well as the rare remains of an early shark. *Hybodus cloacinus* had a dentine spine on its back, here embedded in Jurassic rock. It was found in a Rhaetic bone bed.

WRAPPING AND TRANSPORTING

Most fossils are fairly tough. They need only careful wrapping in newspaper tied with string, to stop them from knocking together in the backpack. Fragmentary samples can be placed in plastic bags; those in soft rocks, such as shale, may need to be packed into a small box with cotton. Labeling the specimens at this stage is important, especially if your expedition has been successful, and you have dozens of finds. Giving locality numbers marked on a topographic map is essential if the fossils are going to have any scientific value.

Some types of fossils are hard, but they are also brittle, and not durable or resilient. They crack or flake at the smallest knock. Therefore, it is wise to protect every worthwhile piece, in case the backpack is dropped or something falls on it.

Strengthening and stabilizing

Some fossils are very delicate: they may be exceptionally thin, or have a flaky, friable surface, or be liable to dry out and crack. The answer is to stabilize and strengthen them in some way before the trip back to the workroom.

The fossils can be painted with suitable glue or resin such as shellac, Alvar or Durafix solution. This serves to strengthen them and stick together any broken parts; it also protects them from drying out, so preventing decay, as well as providing protection on the journey. If you plan to use this method, you should paint it and leave it to dry, after exposing the surface of the fossil, before attempting to remove it from the matrix.

Plaster cast method

For larger fossils, such as a sizeable bone, you may consider making a cast. Make sure you have taken good records of the specimen before you cover it up (see page 60). Cover the exposed and treated part of the fossil's surface with wet tissue paper, followed by several layers of strips of burlap soaked in plaster of Paris. Allow it to set until dry and solid. For very large specimens, professional fossil hunters often strengthen the plaster even more with reinforcing metal rods, since the great weight of the solid stone fossil can easily crack the cast if it is not adequately supported.

A more modern technique is to use polyurethane foam applied over aluminum foil. This again forms a solid, but much lighter, cast. However, the substance that produces the foam needs to be handled with care; it gives off dangerous fumes.

When the strengthening cast is dry, chisel away a bed of rock under the fossil and turn it over. If possible, chip away more of the rock which is underneath the fossil. Then repeat the plastering process on the underside. If the rock is very hard, it may be better to transport the fossil and its bed of rock home in one piece, in a strong box. When the bones of a skeleton are jumbled one on top of another, the whole block should be removed, since separation is rarely possible in the field. You may be able to obtain training as a volunteer at your local natural history museum.

Carbonaceous film fossils

A recently developed technique preserves fossils which may be nothing more than a smear of oily carbon in soft shales. If they are allowed to dry out, they rapidly crumble, and getting them into a collection has been almost impossible in the past.

Brush the loose dust from the surface of the fossil. Pour transparent resin into a collar mold surrounding the outline of the fossil. When it is dry, carefully chip away the block of shale beneath the fossil. Turn it over and clean the underside in the same way. Repeat the resin treatment. You should be left with the most fragile of fossils embedded in a transparent block.

Transportation

Be careful when carrying the packed fossils. If they are too heavy or bulky for your backpack, make several trips. For bigger specimens, you may need to use a wheelbarrow. Get other people to help instead of risking injuring yourself and damaging your discovery. A wooden carrying frame can be constructed around larger finds, if they lack good edges by which they can be gripped and lifted.

It may help to take back extra pieces of rock of the same type as the matrix around your fossil. They can be used in the workroom as "guinea pigs" for testing the various cleaning techniques before you begin working on the matrix itself.

Top and above The plaster of Paris technique, as used by professional dinosaur fossil collectors at the National Dinosaur Museum, Utah. These bones are far too precious to risk damaging during extraction and removal to the laboratory. The exposed bones are coated with burlap soaked in plaster before the underlying rock is cut away, and the block is removed using heavy lifting equipment. This technique was developed in the same area by the teams of fossil collectors led by Marsh and Cope.

To pack and transport the fossil

Once you have found and extracted your fossil, the next stage is to pack it for transportation. For this consult the list below.

- Newspapers and plastic bags for small remains.
- Plaster of Paris and burlap for bigger, more delicate fossils.
- Small plastic boxes, such as margarine tubs, filled with cotton, for carrying small fragments safely.
- You may wish to apply a glue such as Alvar to cement fragments together.
- Labels to identify boxes, bags and packages.
- Cellophane tape and string for general wrapping.

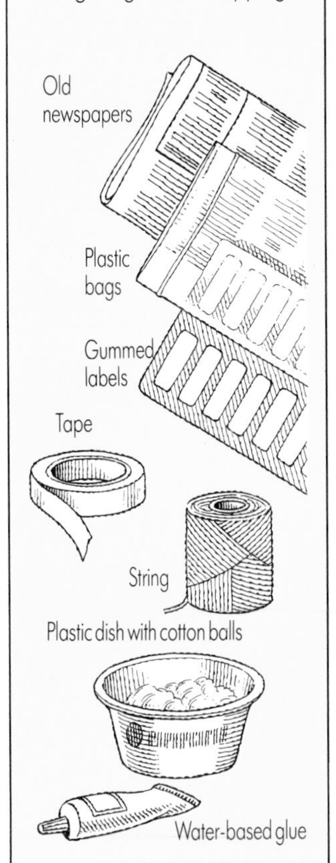

CLEANING AND PREPARATION

Fossils are not cleaned simply to make them look pretty. A well-prepared specimen reveals fine surface details that provide clues such as anchor points for muscles (page 84). In turn, this information helps in interpreting and reconstructing the original organism.

Nature has often started the cleaning process. A fossil is often partly weathered from its rock matrix, which may be the reason why you spotted it in the first place. Your job is to finish the weathering process by hand. The techniques and tools needed vary greatly with the type of fossil and the rock involved. You may wish to experiment and adapt methods to your own needs. Most paleontologists have their favorite methods that they have developed for themselves from the established standards.

If the fossil is harder than the rock which contains it, the cleaning process is usually easier than if it is softer, when chemical application may be needed (page 72).

Washing

Tiny fossils in loose sediments may only need gentle washing in a sieve with water to clean them. They can then be examined under a hand lens, magnifying glass or low-power microscope. Devices are available that deliver ultrasonic vibrations to a small container of water, which loosens the sediments in nooks and crannies, to finish the cleaning process.

Soft rock such as chalk or clay can be washed off under running water with a toothbrush. Be careful not to begin to wear away the details on the fossil's surface. To clean flakes of shale from the surfaces of fossils such as graptolites, deliver a sharp tap with a small hammer and chisel in the direction of the bedding plane. When the fossil is exposed, carefully trim the entire slab to a pleasing shape around the fossil and leave at that.

Larger specimens

When working on a larger fossil, keep it securely clamped to a firm bench or nestled into a sand tray. This limits accidental slips. Support a magnifying glass on a frame above the area on which you are working. If your find is encased in plaster from the journey, it must be carefully cut away with a sharp blade before you can begin to clean away the rock.

If you use a cleaning utensil which is harder than the rock of which the fossil is made, you must be extra careful to avoid damaging the specimen. Use the least destructive implement available, and enlist the rock's help by working along the bedding planes. Point the tool away from the fossil and away from yourself, in case it slips and jumps up.

Slow, patient picking with a mounted needle is

Spondylus spinosus, a bivalve mollusk that lived in warm Cretaceous seas. The soft chalk on the cliffs of England's south coast is easy to clean from hard fossils. This perfect specimen came away cleanly from its matrix, to reveal the characteristic fine radial ribs on the outer surface of the shell.

The fossil hunter's workroom

A fully equipped laboratory or workroom is the dream of most fossil hunters. Unfortunately, finances and household space do not often allow the dream to come true. Nevertheless, it is possible to carry out good work with only a few facilities, together with ingenuity and improvisation.

- A strong, stable, solidly made workbench or table. Fossils are heavy and will put a flimsy table under strain, especially when the hammering begins. Place the bench or table near a window if possible, for natural light, and put a tough floor covering beneath, from which you can easily sweep flakes and shards of rock.
- A work surface such as a small sheet of plywood, which can be replaced as it becomes dented and scratched.
- A sand tray or wooden frame to hold specimens steady while you are working on them.
- A plastic tray into which debris can be swept at the end of the work session.
- A sink with running water; or a large waterproof container with a drain, and buckets for carrying water.
- An adjustable lamp for illuminating the specimen from different angles.
- A selection of small tools including fine forceps, a sharp knife, a hand lens or magnifying glass and mounted needles (of the kind found in biological dissecting kits).
- Various brushes, from fine artist's paintbrushes through toothbrushes to a larger dustpan-type brush.
- A sieve for washing and retaining specimens. Plastic mesh is less harmful to soft fossils, but tends to crack or snap more easily than metal mesh.
- If possible, a frame, or a stand and clamp, to hold a lens for close work.
- A cupboard for storage. If you are using chemicals (page 72), it should be lockable and inaccessible to children.

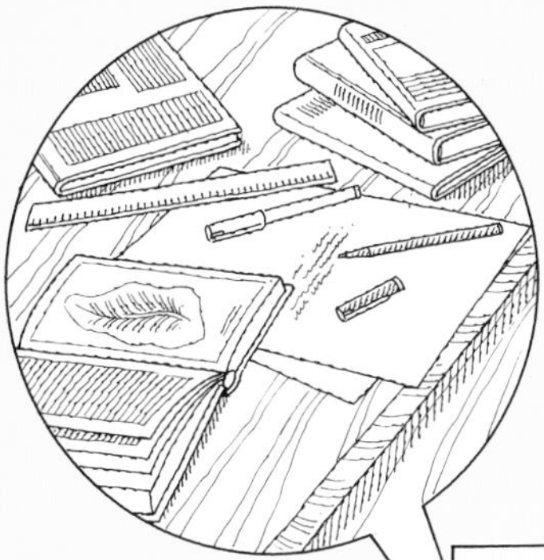

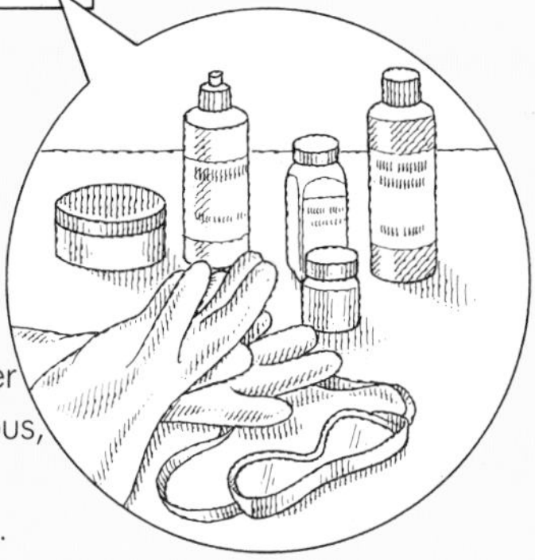

There is an almost endless range of specialist tools for cleaning and preparing fossils available from specialist manufacturers. Your local museum or organization for archaeology, geology, or paleontology should be able to provide manufacturers' details. However, similar tools from hobby shops or hardware stores can be adapted for many jobs. For example, the small hand-held drill used by modelmakers and hobbyists, equipped with a grinding bit, can be used to wear away rock from the specimen.

Left Making drawings and notes, and cataloging the collection, occupies much time. Keep pencils, paper and reference books within easy reach.

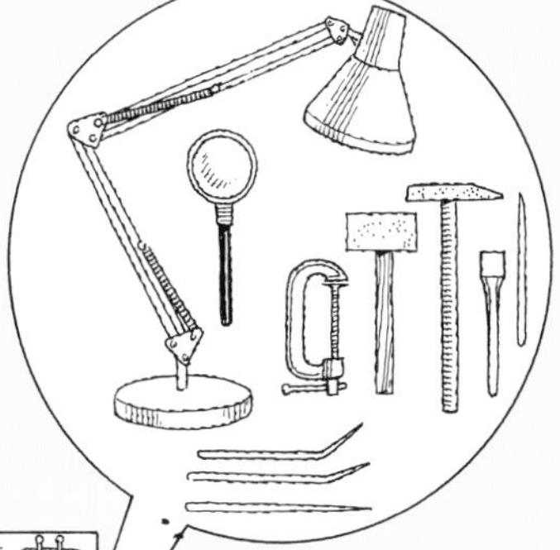

Above Your workbench should be well lit and at a comfortable height. Try to replace tools neatly in racks, and keep your work area free from clutter.

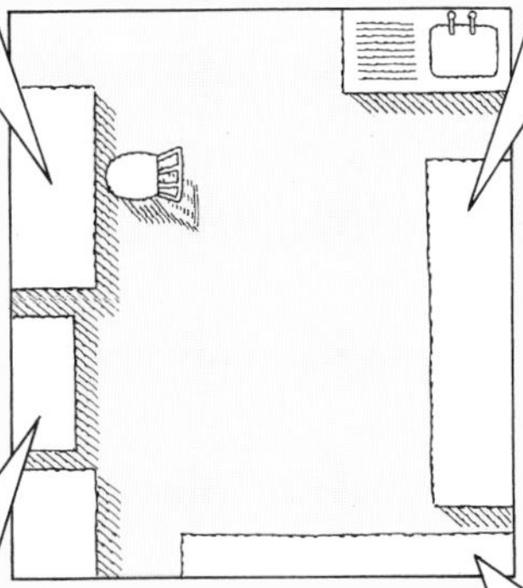

Below You may wish to display your best fossils under glass, while keeping the bulk of your collection in accessible storage. An artist's flat file is ideal.

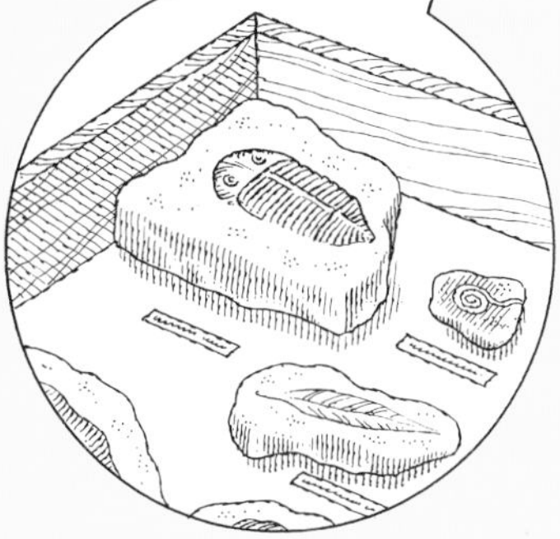

Right Make sure chemicals are stored safely. Bottles must be clearly labeled with their contents and whether they are poisonous, and kept out of reach of children.

the best technique for soft rocks such as shale.

Tough rocks

If the rock is as hard as the fossil, you will need to chip away with great care and patience. You can use a dentist's drill, a pneumatic pick or a hand drill to do this. Specialist tools are often used in laboratories for this process. Small sandblasters which use different grades of abrasive powder are also available. It is wise to get some experience on an unimportant piece of rock of the same hardness before you attempt to clean your fossil. Very delicate structures, such as scale patterns, can be embedded in epoxy resin before the stone is removed from the other side. You might be able to get experience of these techniques by working as a volunteer at your local national history museum.

Chemical extraction

Mechanical extraction of fossils is sometimes not possible. Chemical methods may then be more appropriate; they have the advantage of revealing the finest details, which may be damaged by mechanical methods. Once again, different techniques and chemicals are needed for different types of fossils, and different paleontologists use different techniques.

Clay-based rock can be removed from fossils by boiling them in a weak solution of soda, or heating them to a high temperature in an oven. There are no hard-and-fast rules about time or temperature, so use one of the extra pieces of rock which you brought back from the site to practice.

Microfossils, such as minute teeth, can be separated from loose material by flotation. They will float on certain fluids, such as mineral oil or a strong salt solution, while unwanted debris falls to the bottom.

The skull of *Antrodemus* (*Allosaurus*) being cleaned in the laboratory at the National Dinosaur Museum, Utah. Delicate scraping with fine hand tools is used. The tiny marks and holes on the skull bones, where nerves and blood vessels passed over and through the bone, are vital clues to the reconstruction of the animal. They should not be added to or obliterated by careless work.

Acid techniques

If the chemistry of the matrix is different from that of the fossil, the two can be separated using dilute acids. For instance, if a silicified fossil is embedded in limestone, which is made of calcium, dilute (*not* concentrated) hydrochloric or acetic acid should dissolve the limestone. Phosphate fossils, such as brachiopods or tiny conodonts, are best etched out with acetic or formic acid. Chitinous fossils such as scoleodonts (also microfossils) are extracted with hydrochloric acid. Pollen grains can be etched out using hydrofluoric acid, but this is potentially more hazardous than most chemical methods.

If you are in doubt about which acid to use, try acetic acid diluted to 5 percent with water. Use this solution in a gentle attack on a small, less important area of the fossil.

Acid erosion or etching

When you are treating a fossil embedded in rock, the acid etching or erosion procedure may be suitable. The fossil is covered with a protective layer such as a special glue, and acid is then used to erode the surrounding, unprotected rock surface. It is a lengthy technique with several stages, but worthwhile. Kits are available, containing the relevant chemicals and instructions.

In one variation, the first step is to paint the exposed area with a protective solution of glue made of acetone and amyl acetate (caution: this is highly flammable). Then the fossil is soaked in acid until it begins to emerge from the rock matrix. You then wash the specimen and allow it to dry. Next paint all of the exposed fossil again with glue and soak it in acid until a little more of the fossil is eroded clear. Carefully remove and keep any small fragments of the specimen that may come away during the treatment. Continue the process until the whole fossil is free.

Very fine detail of internal structures or the surfaces of bones can be exposed in this way. Individual bones jumbled together in the rock will eventually fall apart and can be reassembled into a skeleton.

Safety precautions

With all acid techniques, make sure that the fossil is well washed before you start, and again after you have finished. Observe the safety precautions for working with strong chemicals. Acids are corrosive

Etching fossils with acid

One method of cleaning and extracting fossils is by immersing in suitable acids. The technique takes a long time, but you will be rewarded with a fossil undamaged by cleaning tools. Where there are many tiny bones or fragments involved, this may be the only way to retrieve them from the rock.

This is a single, whole fossil embedded in rock. It may be possible to remove much of the rock from around the fossil physically, but the final cleaning is done with acid.

1 The rock bearing the fossil is immersed in the appropriate acid until a portion of the fossil is exposed.

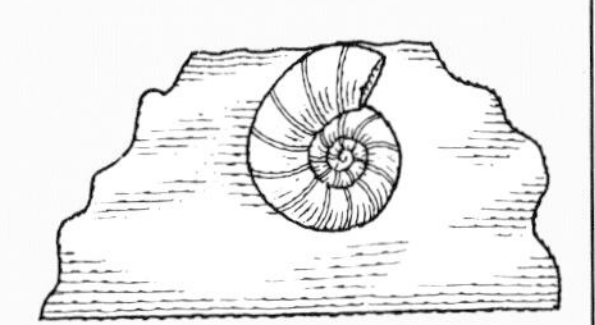

2 The specimen is thoroughly washed and an acid-resistant hardener is applied to the exposed parts of the fossil.

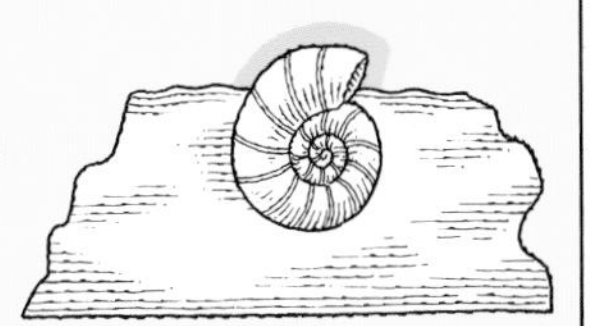

3 The fossil is re-immersed in acid and the whole process is repeated several times over a long period.

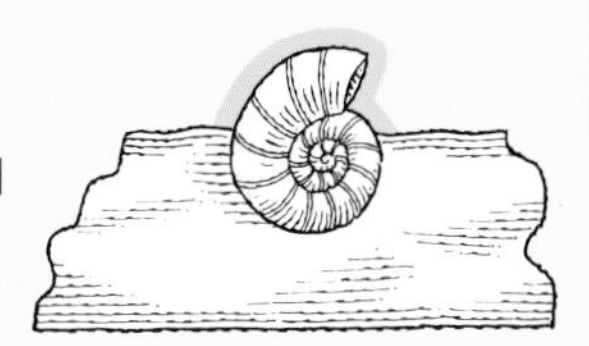

4 Hardener must be applied frequently to prevent damage to the fossil by prolonged exposure to acid.

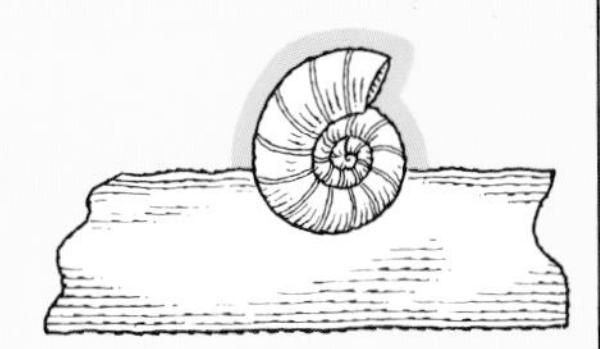

5 Finally the fossil is free of its rock matrix. It should be thoroughly washed to remove the acid and hardener.

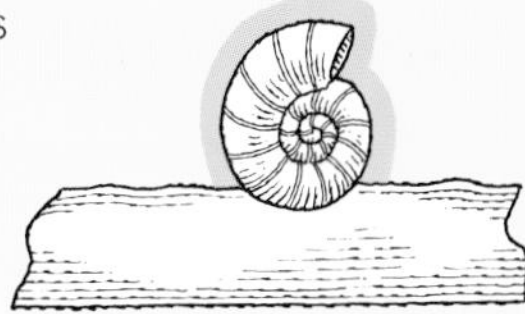

to everything, including the skin, and should be treated with great care. Work in a well-ventilated room, wear gloves and eye protectors, avoid skin contact, and work on a protected surface when using acids. The process can take a long time – from weeks to several years. Do not be tempted to hurry it by increasing the concentration of the acid, or you may lose your fossil and waste all the time you spent in the field finding and extracting it.

Repairing

Breaks in the fossil can be repaired temporarily with water-based glue. A more permanent method is to use resin or a plastic glue, which can be filed and rubbed down to match the shape of the fossil. Sometimes you can mend a fossil after it has been damaged by a break in the chunk of rock that contained it. Stick the whole chunk of rock together again with a resin glue, and then carry out the acid erosion technique as described above. The fossil, already repaired, should be retrieved in one piece.

Molds and casts

A mold is a "shell" of solid material that defines a shape which is the same shape as the fossil. Mold fossils occur naturally (page 33), but you can also make a mold from the fossil itself. A cast is a replica of the fossil, made from the mold. Commercial kits for making molds and casts, usually based on materials such as latex rubber, resins, plaster of Paris and glass fiber, are produced by various manufacturers, such as model specialists.

Making casts of fossils is a technique used by museums for their rarest and most valuable finds. The fossil cast can be put on show to the public, while the actual specimen is kept safely in storage. Paleontologists also study casts when trying to reconstruct the original organisms, instead of using the heavy and precious originals. Indeed, some fossils reveal more detail if the bone is picked away to leave the natural mold made by the rock that has filled spaces within, such as those taken up by the brain, blood vessels and nerves.

Casting is useful when dealing with very fragile, small bones embedded in hard rock. It may be best to dissolve away the bone using an appropriate acid, thus turning the specimen into a mold, and then make a cast of latex from the hole left in the rock. Or casting can be used if the fossil itself has dissolved away during the formation of the rock, leaving an internal or external mold. The result can be as good as, if not better than, the original fossil, since so much fine detail is revealed. The technique is also beneficial when fossils may deteriorate, such as those in iron pyrites.

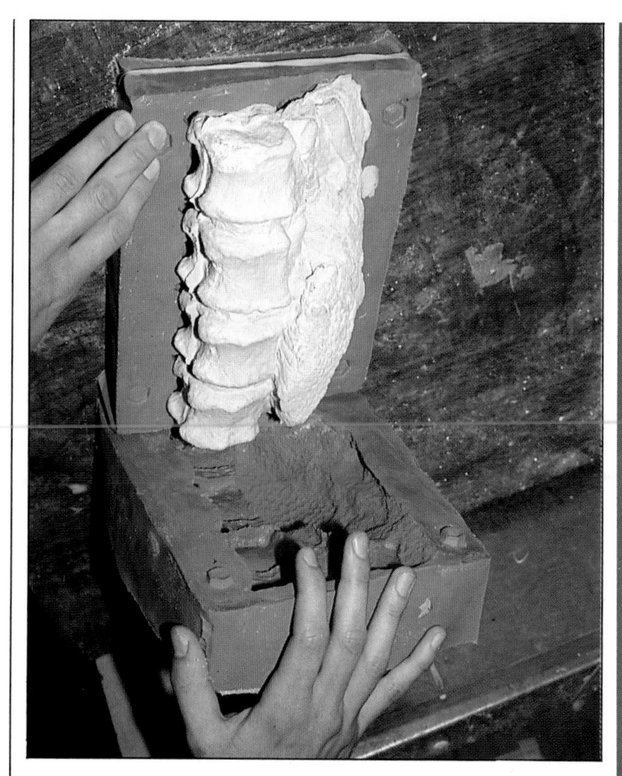

When fossils have been cleaned, they are often cast in plaster so that replicas can be used for display and study. The fossil is encased in a collar which forms a well to hold latex solution. When this is set it can be cut into two halves and removed from the fossil. The mold is then filled with plaster to make as many casts as required.

To reproduce the fossil, make a mold around it by painting on several layers of latex, one after the other. Remove the fossil through a small slit, and make the cast with plaster of Paris.

If you already have a natural mold fossil, make a cast from it in the reverse way. Modeling clay is a rough-and-ready material; for finer detail, latex rubber is best. Clean the mold well. The flexible latex can be used to make casts of the most complicated mold, yet is still relatively easy to extract.

Above and below Trilobite fossils are common in Paleozoic marine deposits. This is *Diacalymene drummuckensis*, an Ordovician species. In fact, these fossils are usually not of the whole animal, but are casts of the molted exoskeleton, shed as the animal grew. The connections between the parts of the "skins" were so strong that they often curled up after being shed. *Flexicalymene retorsa* is also an Ordovician trilobite that could roll up.

IDENTIFYING FOSSILS

Identification of fossils gives your collection an added dimension, in addition to the purely aesthetic quality of the specimens as objects. It also encourages you to look carefully and closely at each fossil, to find the key identifying features, thereby helping you to appreciate and understand your finds more completely.

It may seem a daunting prospect at first. To the unpracticed eye, one amonite looks much like another. However, most enthusiastic beginners quickly become more expert as they go.

The fossil hunter's library

Reference books are essential: both fossil identification guides and general biology textbooks. You do not have to buy them; use local libraries, universities and museums. Sometimes, guide books specific to the fossils found in an area are published by a major local museum. General fossil books illustrate examples of each main group of animals and plants, for comparison. They may also contain keys for identification.

Museum or university fossil collections are also invaluable. If you still cannot identify your find, there is usually a member of the staff willing to help the genuinely interested amateur. If your find does turn out to be rare and important, they may ask you to donate it to their collection.

Order in the natural world

Some knowledge of the science of taxonomy, classifying animals and plants, is needed to understand what your fossils are, and how they are related to other groups, living and long extinct. The chart on pages 112-113 provides a broad outline of the main fossil and living groups.

The rules for identifying a fossil are much the same as those for living organisms. Use the keys found in the specialist books, which can often take you from phyla (the major groups) to smaller groupings, such as genera for invertebrates. Vertebrates are more difficult, because you may be working with small fragments, like bits of bones or teeth, rather than "whole" organisms such as complete shells.

Circumstantial evidence

The locality and stratum in which the fossil was discovered, and the other fossils lodging with it, are often strong clues as to its identity. A specimen in rocks 300 million years old, for example, is unlikely to be from a mammal, since mammals probably appeared about 80 million years later.

Fossil populations cover long periods of time. Within this, there is variation between individuals,

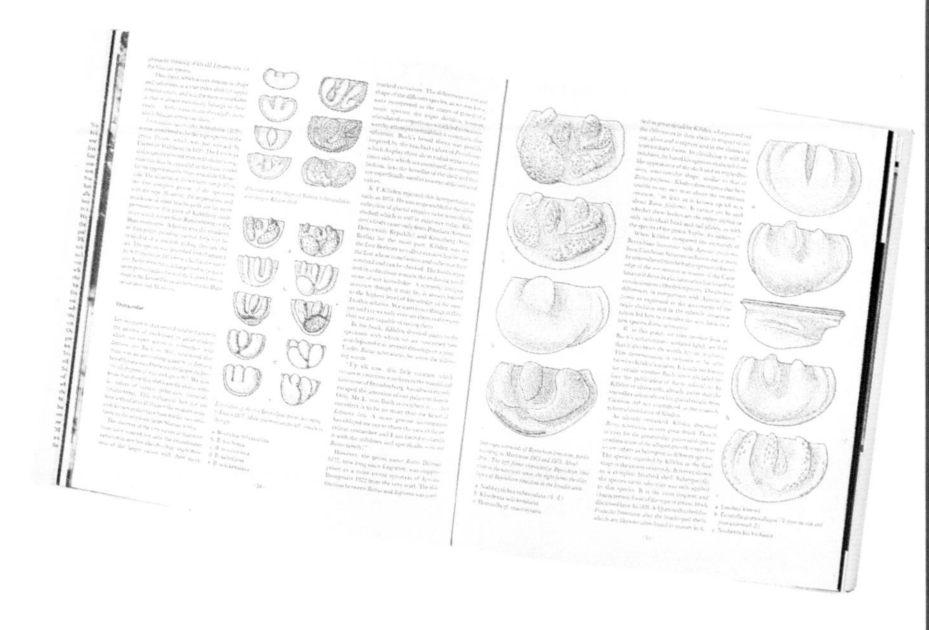

A detailed reference book is essential for the study of fossils. Choose one that covers your chosen field. Specimens, even of the same species, vary considerably because of chronological and geographical separation. These variations can be confusing if you don't have good reference material – but they may lead you to new information if carefully recorded.

as there is in animal and plant populations today. It is often difficult to say whether a fossil is a new species or a variant of a known species. Misinterpretations have been made throughout the history of paleontology, and specimens are constantly being re-studied in the light of new evidence, theories and techniques. Indeed, as fossils change slowly over time, from one type to another, it is often difficult to say where one species ends and a new one starts.

Two of the very rare and famous fossils of *Archaeopteryx* (the earliest bird found so far), were at first identified as a small dinosaur and a pterosaur (flying reptile) respectively. Only when they were re-examined many years later was their significance realized.

Working through the taxonomic groups

Having decided that your fossil is not a pseudofossil (page 35), first assign it to a kingdom: is it plant or animal? This initial decision may not be as easy as it sounds, since many animals resemble plants if there are no clues as to their color.

The initial hurdle over, assigning a phylum – such as fern, flowering plant, coral, echinoderm, mollusk, arthropod or vertebrate – is usually less of a problem. It should not be too difficult to decide on the next two levels, class and order, especially for invertebrate animals. Progressing much beyond this, to name the genus and species, tends to be the realm of the academic experts, who themselves often have problems. Your fossil may simply not have the essential part that allows you to name it.

Left Carboniferous crinoids carpeted the seabed of what is now Derbyshire, England. On their death, the external skeletons broke down to the component ossicles. The shapes of the ossicles indicate the species of crinoid. Undamaged, these calcareous remains became rocks, known as crinoidal limestone. The study of the deposition and interrelationships of animals in such assemblages is known as taphonomy.

Above Identification of plant remains is often difficult. This leaf is preserved in Eocene sediments. Identification often depends on comparison with living species. The more recent the fossil, the more useful this technique is. This is probably a laurel leaf.

MAKING A COLLECTION

Fossils are generally easier to keep than to find. They do not deteriorate, except in rare cases; their main needs are space, an occasional dusting and some appreciation.

As a beginner, you probably already have some idea of whether, or how, you wish to keep and display your finds. Some fossil hunters love the searching and field work, while others put up with those aspects of the pursuit in order to enjoy their well-presented, finely housed, neatly labeled collection.

Some collectors embed fossils in clear plastic. Others varnish their finds. In general, do as little as possible in order to keep the specimens in a stable condition. Varnish may give the rock a gloss or sheen, but it obscures fine detail and is difficult to remove. Fine details are important, particularly if your collection is to form a basis for study, so that eventually you will be able to make some contribution to the body of paleontological knowledge. Each fossil, even if only a fragment of the whole, is unique and worth treasuring.

Presentation

Give some thought to how you present each fossil. Small specimens lie neatly in rows. Pieces of rock bearing traces of fossils can be propped up on wooden supports. Fossil bones often look interesting left partly exposed in a neatly trimmed block of matrix.

If you have managed to extract several bones from what you assume to be a single skeleton, you may wish to reconstruct it (page 82). This requires a good knowledge of zoology, and reference books are again necessary. If you manage to piece the parts together, hold them in place with fine, soft wire, and mount the skeleton on a baseboard (page 93). You can try modeling the missing pieces in clay, but they should stand out clearly as artificial.

Labeling and information

If your collection is to have any real use, the fossils should be carefully labeled with all the information you detailed when you found them. Note the date of the find and who made it; the exact location of the find; the name of the location or a map reference, together with some sort of reference point, such as a tree or a telegraph pole. Even more important is the type and age of the rock bed which yielded the fossil. If you are not sure of the exact age of the stratum, record details such as its color

"Iron pyrites disease"

During the last century, when the major museums began to collect dinosaur fossils in great quantity, some of the remains started to suffer from a condition called "iron pyrites disease." Its cause was thought to be bacteria, which made the fossil decay and turn to dust. It is known that a chemical reaction occurs, as the pyrite slowly reacts with the moisture in the air, and feathery crystals begin to grow on the fossil. Antibiotics have been used against the problem in the past, but prevention of desiccation is the answer. The process can be slowed by exposing the fossil to ammonia vapor; the fossil is then stored in silicone-based preservative, or even glycerine, in an airtight jar.

The Jurassic ammonite *Dactylioceras*, prepared for display. It has been sectioned by grinding to reveal the inner whorls and the septa between the shell's compartments that formed as the animal grew.

Mollusks form a good basis for a collection. They are found in most marine and freshwater sediments, and there are a huge variety. They are conveniently sized to display attractively, and it is possible to specialize. These are Eocene examples: a gastropod *Turritella* (center), and bivalves *Cardiocardita* (left) and *Charma* (right).

Fusus antiqua var. *contraria*, from the Pliocene/Pleistocene border in Suffolk, England. Most shells have sinistral coils; this is dextral.

Cerithium duplex, a gastropod from upper Eocene rocks. The decorative spikes and shape of the opening aid identification.

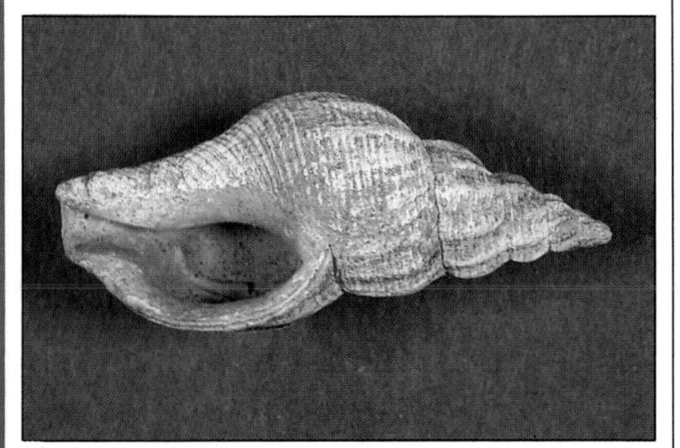

Searlisea costifer, from Pliocene coralline. This animal still lives in the North Atlantic.

Pleurotomaria bitorquata, a Jurassic gastropod. This primitive snail is similar to the modern top-shell.

and grain size, and similar information on the rocks above and below it. Also write down any thoughts you have about the life or death of the organism, or other fossils found in the same area.

The label should also give details about your identification and interpretation of the fossil. All organisms, living or extinct, are classified according to a system devised by the naturalist Carolus Linnaeus in 1753. Each organism is given a scientific name (usually Latin-derived) with two parts: the name of the genus, or group of closely related organisms; followed by the species name. The scientific name is always written in italics or underlined, with a capital letter for the genus name only. If you have several fossils belonging to the same genus, you can use the initial letter of the genus name only. If you are unsure of the species name, you can replace it with the letters "sp."

Cataloging

As this information is too much to write on a small label, you will need to set up a cataloging system. With this system, the fossils themselves are labeled with a reference number written in black or white ink, whichever is most appropriate, in a place which does not interfere with its inspection. Remember that gummed labels have a habit of falling off, especially when they become old and dry. The catalog lists all the reference numbers, along with the information relevant to each fossil.

You may also wish to include details of the cleaning technique used, so that you can repeat the process – or learn by your mistakes! Try to keep your catalog up to date, since the collection could soon overtake your memory.

Cases and containers

Your fossils will be of various sizes, and some will be more delicate than others. How you store and display your collection depends on your particular interest, but the main requirement is that the fossils should be protected from dust and mechanical damage, yet remain easily accessible for examination.

It is possible to buy cardboard boxes which fit together neatly side by side, no matter how many of each size are needed, without wasting space. Lined with soft cotton batting, these boxes can be fitted into a narrow drawer such as a flat-file drawer, enabling the fossils to be stored neatly in a dustproof manner, but still easily accessible.

You may wish to arrange the fossils in their drawers group by group, perhaps with the "lower orders" such as sponges at the bottom. Or the

Below and right A typical amateur collection, with the specimens simply arranged for best visual effect.

The owner worked hard to find and prepare the collection, and has every right to be proud of it. Attractively displayed, a home collection holds a fascination for others, be they fossil experts or not.

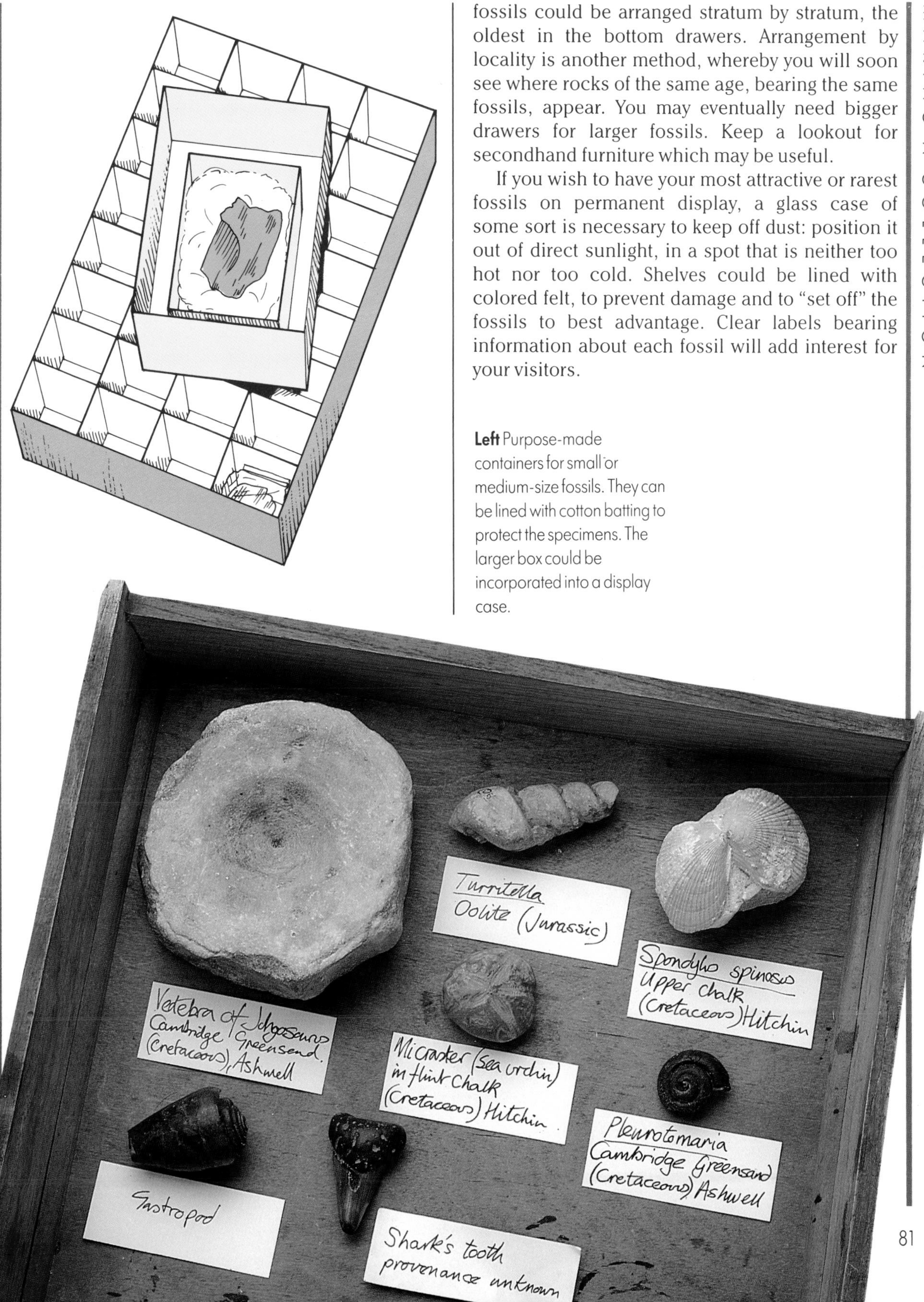

fossils could be arranged stratum by stratum, the oldest in the bottom drawers. Arrangement by locality is another method, whereby you will soon see where rocks of the same age, bearing the same fossils, appear. You may eventually need bigger drawers for larger fossils. Keep a lookout for secondhand furniture which may be useful.

If you wish to have your most attractive or rarest fossils on permanent display, a glass case of some sort is necessary to keep off dust: position it out of direct sunlight, in a spot that is neither too hot nor too cold. Shelves could be lined with colored felt, to prevent damage and to "set off" the fossils to best advantage. Clear labels bearing information about each fossil will add interest for your visitors.

Left Purpose-made containers for small or medium-size fossils. They can be lined with cotton batting to protect the specimens. The larger box could be incorporated into a display case.

INTERPRETATION AND RECONSTRUCTION

Since the early days of paleontology, collectors have tried to piece together the parts of their finds, and the evidence contained in them. This would allow them to reconstruct the complete organism as it would have been in life. They have had varying degrees of success in this. Baron Cuvier was an early master of the technique, and his basis was a sound knowledge of comparative anatomy.

In vertebrates, and especially in tetrapods (four-legged vertebrates from frogs to giraffes), the bones of the skeleton follow the same general pattern. If you know the structure and function of a certain bone in one animal, it is possible to estimate the shape and working of the equivalent bone in another, similar animal. The more closely related the two animals are, the more accurate this process becomes.

The work falls into two main areas. The first is interpretation. This involves close examination of the fossils, noting features such as areas for muscle attachment, or articulations where one bone was linked to another at a joint. This leads to proposals concerning the organism's anatomy, its way of moving, its diet, etc.

The second stage is to build on this knowledge by reconstructing the skeleton, and even the whole animal or plant, as it appeared in life. This may be highly conjectural, since dinosaur bones, for example, do not give us any clues about the color and patterning of the animal's skin.

Each fossil is unique, and a book such as this can only provide general guidance. The diagrams and photographs in this chapter show specific examples. If you have a prize fossil that you wish to describe and reconstruct, consult books that specialize in the paleontology and anatomy of the relevant taxonomic group.

Early reconstructions

The great dinosaur models built by the sculptor Benjamin Waterhouse Hawkins and Richard Owen (page 17) during the 1850s were some of the first attempts at reconstruction. They pictured the animals mainly as overgrown frogs or turtles. Hawkins was invited to New York City to recreate his models for the Paleozoic Museum, but the city found it could not justify the expense involved; some thought the models heretical, and workers smashed them to bits. Gideon Mantell's original interpretation of *Iguanodon*, as a beast on all fours with a spike on its nose, is another famous early reconstruction.

Reconstructing evolution

It may be possible, if the fossil record is complete enough, to reconstruct the probable evolutionary development of organisms. Each stage shows the slight changes that have taken place as one species is replaced by another. The development of the horse is one famous example; it has been traced from the early Eocene *Hyracotherium* (named by Richard Owen) to the present-day species. Similar sequences have been made for rhinoceros and elephants, which include the mammoths of the cold Quaternary climate. Perhaps the most important and hotly debated evolutionary reconstructions are those of our own ancestors.

Fossils can never prove that one species was the true ancestor or descendant of another; only going back in time and observing the process in ac-

Cladistic analysis

Relationships between animals can be illustrated by a branching diagram called a *cladogram*. A group of animals that share a common ancestor is called a *clade*. Animals that share a common characteristic or homology may belong to a clade, and are therefore more closely related to each other than to animals which do not share the homology. Problems arise where animals have lost certain characteristics during evolution: then the relationships are less obvious. For example, birds have lost the hand claws, long tail and teeth which they probably once shared with the dinosaurs. Sometimes the same features have appeared independently in separate groups which have had to adapt to similar habitats, an example being the flight surfaces of bird wings, bat wings and pterosaur wings (see page 89).

A
B
C

Cladogram showing **A** most closely related to **B**, and both less closely related to **C**.

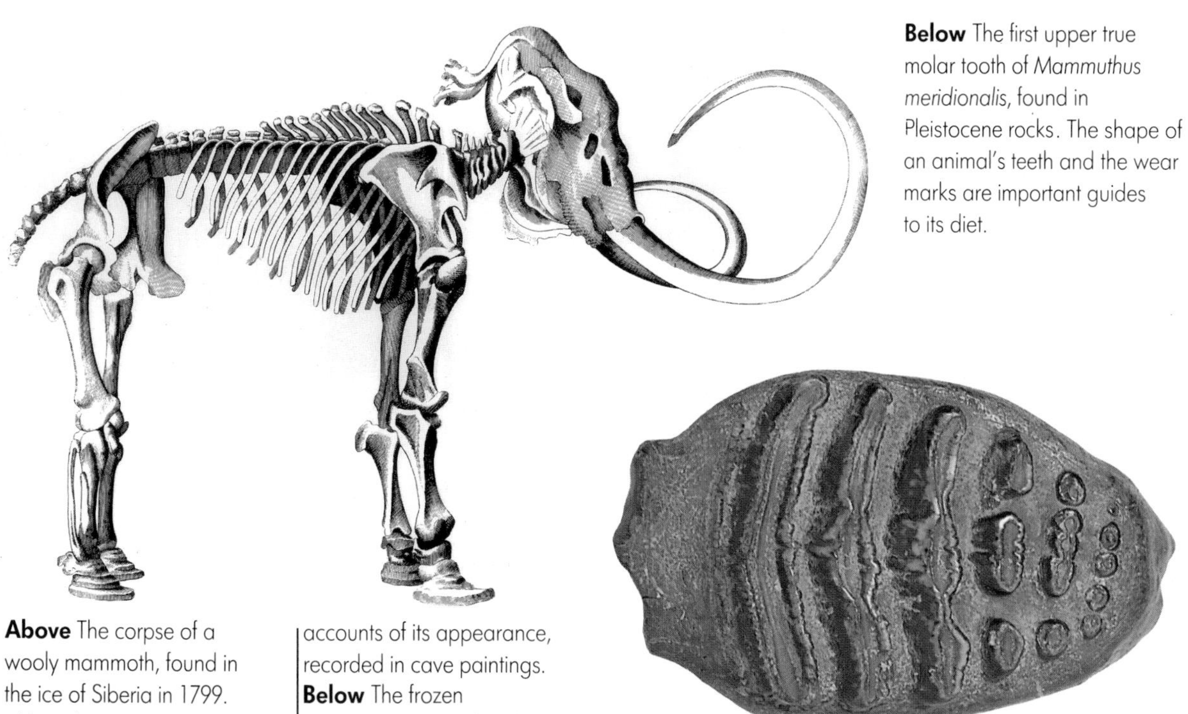

Below The first upper true molar tooth of *Mammuthus meridionalis*, found in Pleistocene rocks. The shape of an animal's teeth and the wear marks are important guides to its diet.

Above The corpse of a wooly mammoth, found in the ice of Siberia in 1799. Although extinct, its reconstruction was straightforward because complete specimens were preserved, deep-frozen. It is also one of the few prehistoric animals for which there are eyewitness accounts of its appearance, recorded in cave paintings.

Below The frozen mammoths, studied and reconstructed by Cuvier, excited popular imagination. Artists sketched the scene of the discovery since there were no cameras at the time.

tion could show that. But the probability that one species is descended from another, very similar species in preceding timespan, is more than likely.

Drawing on other disciplines

To obtain the maximum information from fossils, the modern science of paleontology has many years of experience behind it, and it borrows techniques and expertise from many other disciplines. These techniques are constantly being modified and new ones invented.

Clues are gleaned from the fossil itself, with both external and internal features if possible. They are also gathered from the rock stratum in which the fossil was found, the age, the type of sediment, and any other fossils around it. This evidence is weighed against and compared with our knowledge of living organisms and their habits.

Today, thousands of professional paleontologists labor to populate the ancient seas with fearsome swimming reptiles, or to litter the land with action-packed scenarios of the first amphibians crawling out of the water, or dinosaurs hunting in packs. Many museum exhibits are no longer filled with inanimate, dusty bones. They consist of modeled reconstructions that move, roar, fight and protect their young against realistic backgrounds.

CLUES ON AND IN THE FOSSIL

When studying vertebrates, from fish to mammals, you will notice that the bones of the skeleton follow the same basic pattern. The main problem is the initial decision of bone identification.

The bones of an animal tell us how it moved. The stresses on bones and joints are different if the animal is four-legged or a biped, or if it moved fast or slowly. These stresses change the strength and reinforcements of the bone and the way it articulates with the bones around it. Bones also bear scars where the muscles were attached, and the sizes of the muscles can be estimated according to how much weight they had to move. So, to an extent, we can put flesh on the bones.

Occasionally, skin has left impressions in fine sediments. Some dinosaur skin, for example, contained bony plates, while some pterosaurs had hair, and *Archaeopteryx* had its feathers. The color of the skin remains a mystery, but scientists and artists can consider what we know about the creature's lifestyle, and the reasons for animal coloration today, to make informed guesses.

Above The humerus (upper foreleg bone) of *Platypterygius campylodon*; a Cretaceous vertebrate. The shape of this important bone indicates the type of foreleg, and therefore how the animal moved. Close examination may reveal muscle scars.

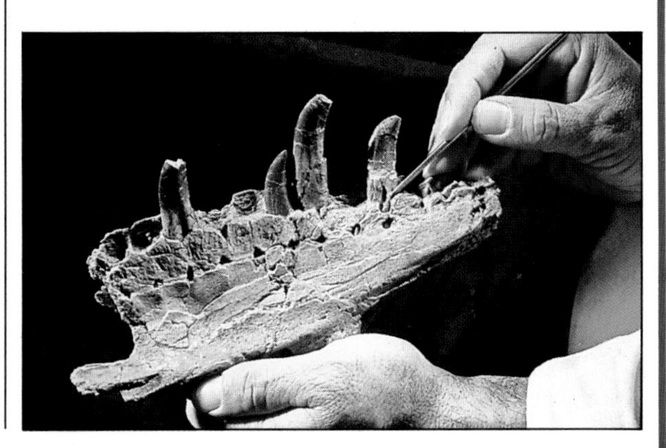

Right The jaw of *Antrodemus* (*Allosaurus*), a large carnivorous dinosaur. The long, curved, serrated teeth and deep jawbone indicate how the animal cut and tore flesh from its prey. The teeth were replaced when they were worn or lost.

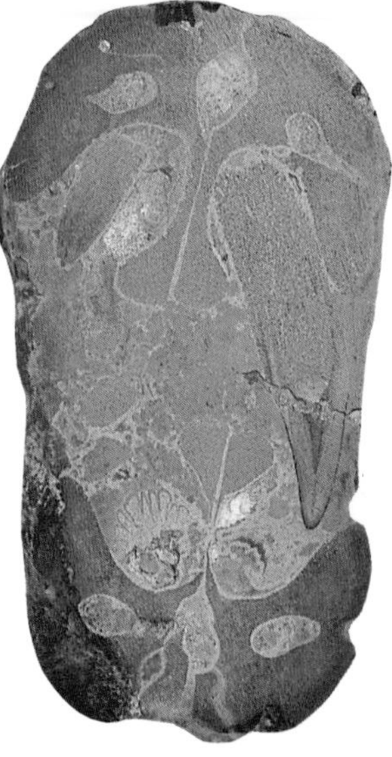

Far left The jaws of the large Jurassic swimming reptile, *Ichthyosaurus*. Seen from the side, the teeth are those of a fish-eater – the long, thin jaws are ideal for catching fish as they swim by. This fossil has been sectioned across the jaw. **Left** From the front, it reveals much more about the anatomy of the animal. The structure of the teeth and how they are set into the jawbone, the shape of their roots, and the cavities within the bones are all clues.

Teeth

Teeth give many clues as to what the animal ate. Are they sharp carnivorous teeth for piercing and tearing flesh, or blunter herbivorous teeth for slicing and crushing vegetation? Small peg-like teeth are often for fish-eating (known as piscivorous dentition), and fine straining structures are for filter-feeding.

Wear marks on the teeth indicate the direction of chewing, if any. Together with details of the articulation and musculature of the jaw, this evidence can give a good idea of the animal's diet.

Invertebrates and plants

Most amateur fossil hunters come upon the remains of invertebrates and plants, rather than those of the large and spectacular vertebrates. When studying the former, there have been so many variations over the millennia that it is sometimes difficult to know where a fossil fits. A small fossil may be damaged, or the crucial piece might be missing.

The classification of plants depends on the different types of seed-bearing parts such as flowers, and the leaves, bark, roots, internal structure and method of germination. If your fossil consists of a leaf fragment or a bark cast, reconstruction may be next to impossible. Comparison with fossils already found in the same area and strata offers one solution.

Invertebrates are usually – if not completely – formed from soft tissue, which is rarely preserved. However, scars on the inside of a shell sometimes indicate where the animal's muscles or breathing apparatus was. It is only from the rare finds of soft-tissue impressions, or the beautiful X-ray photographs of inner details, that paleontologists can deduce the internal structure of these creatures and how they lived.

Deductions about lifestyle

The overall shape of an animal with an external skeleton, such as an insect or echinoderm, may have been deformed by the processes of fossilization and subsequent rock movements. But the general form provides clues as to where and how it lived. If it had a "stalk," such as a crinoid, it probably lived on the bottom of the sea. Casts of different trilobites have been tested in water currents; their hydrodynamic properties indicate that some lived on the seabed while others were active swimmers.

The surface features of the fossil may also give clues as to whether the organism needed to defend itself from some fearsome carnivore, or perhaps it suffered bite and chew marks as it met its end.

The coral *Acervularia ananas*, from Devonian rocks. These types of coral are responsible for reef-building in the past, as now. The anemone-like animal secreted a calcareous cup around itself for support and protection. Each animal is usually part of a large colony. Over thousands of years, millions of cups build large mounds of calcareous rock. Corals have distinctive radial patterns for identification, and they are good indicators of conditions such as depth and temperature, since they are choosy about their environment.

CLUES FROM AROUND THE FOSSIL

Tracks, trails, nests and feces (together known as trace fossils, page 33), combined with information from the site where the fossil was found, all yield more clues. They can be assembled to build up a clearer picture of the fossil in life. All this evidence comes from around the fossil, not from the fossil itself, and it illustrates the importance of careful site recording (page 60), and logging details of fossils found together.

Tracks, trails and prints

Tracks and burrows are often the only evidence we have that some invertebrates existed. Huge numbers of worm-like animals from several major groups are only known from the trails they left behind or the burrows they built for themselves in rocks. Furrowed tracks of trilobites, the groove made by a rolling ammonite, and trails of a giant centipede six feet long, may tell a small story about how the creatures lived and died.

Footprints can be studied to discover how heavy the animal was, how fast it was moving, and whether it traveled alone or as part of a herd.

Fossilized food

Fossil feces, or coprolites, are found in great numbers in some areas. If you know which animal produced them, their contents suggest what it ate. Sometimes scattered near a fossil, especially a vertebrate, are remains from its stomach – evidence of the animal's last meal.

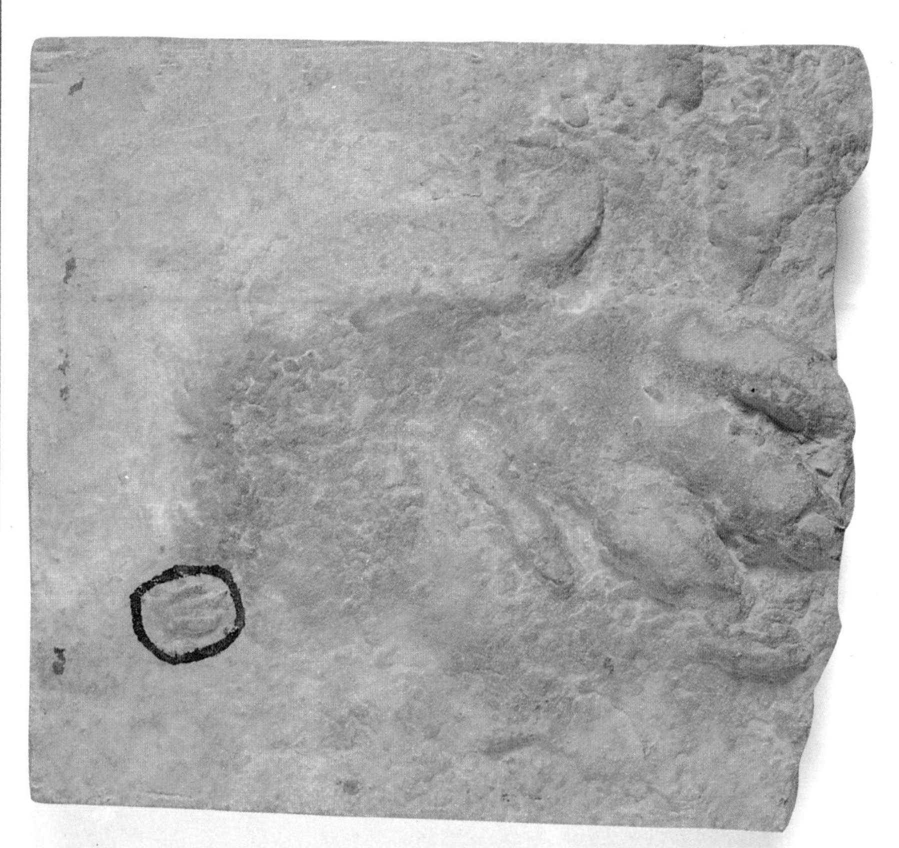

Above A coprolite found in Lower Lias Jurassic rocks. It was probably produced by a large marine reptile, such as a plesiosaur, fossils of which also occur in these rocks. Although it looks like any other rock, the shape and swirling patterns on the surface of a dropping are often very characteristic of the animal that left it.
Left Footprints in the sand, usually so fleeting, are on rare occasions preserved for millions of years. This specimen of a labyrinthodont amphibian was found in sandstone rocks. It is probably about 300 million years old. Note the much smaller print inside the dark circle.

When a number of fossil plants and animals from the same era and region have been reconstructed, a skilled artist can bring them to life. This is a Carboniferous scene.

Breeding and social life

As the soft organs are very rarely preserved, details of how animals behaved and reproduced are sketchy. However, with recent discoveries of dinosaur nests and eggs, these details are becoming clearer. It is now believed that some dinosaurs were very sociable animals, they probably communicated vocally, and they were good parents. The eggs often still contain the embryos, which can be seen by the CAT (computerized axial tomography) scanning methods used in hospitals.

Plants

Plant fossils are much easier to identify if their seeds and pollen can be extracted from the surrounding rock. Both tell a great deal about how the plant grew and reproduced. Pollen grains are very characteristic; an entire branch of paleontology is devoted to their study. They can also act as indicators to date other fossils in the rock.

Fossil plant stems tell us about the way plants grew at the time. Was it a dry-land forest, a grassland, or a swamp? If a woody trunk has been preserved by silica, the fine detail of the annual growth rings may be as good as those of living trees for dating and climate analysis. The trunks of tree ferns contained no true wood, and only molds or casts of the bark are preserved. The pulp inside rotted before the bark, and some specimens contain an added bonus: remains of animals that were trapped or took refuge inside. A very early reptile, *Hylonomus* of 300 million years ago, was preserved in this way.

Information from the strata

The rock stratum in which the fossil was found is the best clue to when the specimen lived. The layer can be carefully dated by its indicator fossils. The type of rock also suggests where the animal died and how it was fossilized.

But there are pitfalls. Be on your guard when looking at the rock strata, because animals are sometimes fossilized when they fall down deep holes into older strata, and are rapidly buried. If the type of rock immediately around the fossil is different from that of the main rock face, this may be what happened. You therefore need to track upward (usually) to the younger rocks to locate the true position of the sediments in which the creature was entombed.

COMPARISONS WITH LIVING FORMS

The main clues for reconstruction come from comparisons with living animals and plants. The rules that govern how animals and plants cope with life are presumably the same today as they have always been. Some of the organisms that lived on Earth left no descendants, but many are still represented. These "living fossils" can show how long-dead organisms looked and functioned. Fossils of animals and plants that have no modern counterparts are the most puzzling. *Hallucigenia* was a strange appendaged animal found in the famous Burgess Shales, so named because those who examined it thought they were seeing things.

Convergent evolution

Sometimes, similar but unrelated forms can give clues for reconstruction. The streamlined body and the fin or flipper arrangements of modern aquatic animals, from fish to seals and whales, tell us a great deal about fossil aquatic animals: at what depth they lived, how fast and maneuverable they were, and how they caught their food.

Convergent evolution occurs when similar environmental pressures are placed on organisms. The "design solutions" are often strikingly similar, even though the animals are not closely related. There are many examples: one of the best is the comparison of modern dolphins (which are mammals) to the ancient reptilian ichthyosaurs, such as *Shonisaurus* from Ichthyosaur State Park in Nevada. Both groups of animals are superbly adapted to ocean life, and both evolved from air-breathing terrestrial animals. Their overall appearance is very similar: both possess long beak-like mouths lined with peg teeth, and their paddles have the same arrangement. Only the tail is different. That of the dolphin sweeps up and down; being a mammal, its muscles and bones are arranged in the typical mammalian pattern, and you may imagine similar undulating movements in the body of a cheetah at full speed. The reptilian

Crocodilians first appeared over 200 million years ago and have changed very little since. Compare a version of one of the first crocodile ancestors, a reconstructed *Protosuchus* from 200 million years ago (left), with the living species of today (above).

pattern allows side-to-side sweeps, as we see in today's crocodiles; the ichthyosaur's tail moved this way.

In the air

Like swimming in water, flying in air places certain demands on the shape of an animal. Several groups have tried flight; the insects have been aloft longest. They use mainly flapping flight, with muscles capable of beating the wings consistently and very fast. This limits the insects' size. Flying insects have stayed small, the largest being dragonflies such as *Meganeura* from the Carboniferous coal forests, with a wingspan of 26 in. (65 cm).

Several groups of vertebrates have tackled flight. On the whole, they are bigger and heavier than insects, so sustained flapping is more strenuous. Gliding is much more a part of their flight. Several reptile groups, including the pterosaurs, mastered the skies for over 100 million years during the Mesozoic era. They were very successful and produced the largest animal that ever flew, Quetzalcoatlus, with a wingspan of 40 ft (12 m).

The pterosaur wing is similar in outward shape to that of birds and bats, but it is arranged in a different way. A flap of skin is stretched out from the greatly elongated fourth finger, along the arm and down the side of the body. The bird's wing is held out by the arm and wrist bones, with the fingers supporting only the region near the tip; feathers replace the skin flap. The bat's wing is again a skin membrane, but it is supported internally by four elongated fingers.

These comparisons illustrate the lines of reasoning used in interpretation. If a new fossil creature with wing-shaped forelimbs is discovered tomorrow, we would at once assume it was a flyer. The structure of the body and wing would place it into one main group: reptile, bird or bat. From there, comparisons with related animals lead along the road to reconstruction.

Weaving a conclusion

Fossil spiders have been found in Sierra de Montsech, in northeastern Spain, in limestone 138 million years old. The fine details show both their spinnerets, for spinning the thread, and the claws on the ends of their legs, used for weaving the threads into a web. Modern spiders that weave webs have the same arrangement of claws. Previously it was known that spiders from as far back as 300 million years ago spun thread, but not that the threads were woven into webs.

This simple example shows how comparison with living forms can enlighten us about the habits of fossilized forms. The webs themselves have long gone, but we can presume the spiders made them. Otherwise, why would they have clawed legs?

Convergent evolution

When different animals have to deal with the same environmental or lifestyle problems during their evolution, they often develop similar solutions. Pterosaurs, birds and bats, for example, have all conquered the air by developing wings and the ability to fly. Wings have arisen independently in these animals, and are formed by different arrangements of the bones in the basic tetrapod limb.

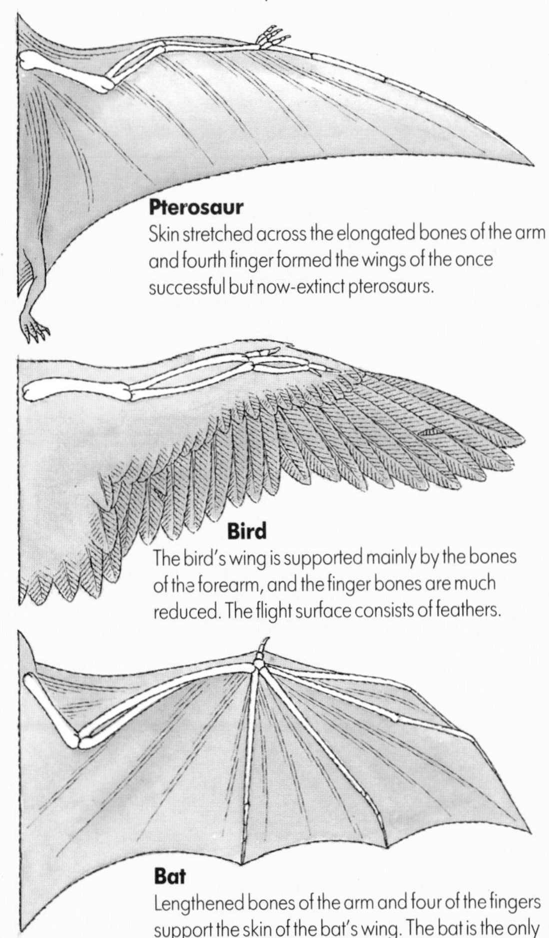

Pterosaur
Skin stretched across the elongated bones of the arm and fourth finger formed the wings of the once successful but now-extinct pterosaurs.

Bird
The bird's wing is supported mainly by the bones of the forearm, and the finger bones are much reduced. The flight surface consists of feathers.

Bat
Lengthened bones of the arm and four of the fingers support the skin of the bat's wing. The bat is the only true flying mammal.

PROFESSIONAL RECONSTRUCTIONS

Reconstructions need to be as perfect as possible to allow scientific study. Fossils in museum collections are studied by paleobiologists from all over the world. The information is recorded and published in the scientific literature, so that it can be interpreted and used by other scientists. A full description and an illustration or model of the reconstructed organism finalizes work which has probably taken many years.

The techniques used by the professionals are borrowed from medicine, chemistry, electron microscopy and electronics. All require expensive machinery, but they offer interesting insights for the amateur fossil reconstructor.

X-rays

X-ray photographs can be taken through thin slabs of rock, revealing the finest soft-tissue details of invertebrate animals. Using this technique, such animals as belemnites, previously known only by their single internal pencil-like shell and a few incomplete impressions, can be seen in almost living detail.

Larger skeletons, such as the big reptiles of Triassic bituminous schist in Switzerland, have been photographed without extracting them from the fragile rock. CAT scans can reveal the inner contours of solid fossils such as skulls; the size and shape of the brain then gives much information about the behavior of the animal.

Microscopes and cameras

Microfossils can be viewed under the microscope in very thin sections of rock cut with special machinery. Both light and electron microscopes are used on thin slices or sections of fossil bones, revealing internal details often almost identical to modern bone. Reptile bone can be distinguished from mammalian bone in this way.

Photography under infrared or ultraviolet light can reveal details invisible in ordinary light. Detailed drawings and photographs of the specimen are taken from many views. Stereo three-dimensional photographs of complicated structures such as teeth may be easier to study than the actual fossil. These visual records are accompanied by written details, in strictly defined anatomical language.

Reconstructing the organism

Theories about how an organism looked or lived can be tested with a modeled reconstruction. When the enormous pterosaur *Pteranodon* was first discovered, it seemed unlikely that it could fly. It had a long beak at the front of its skull and a long counterweight projection behind its head; there was no stabilizing tail; its legs were short; and its size was startling. But surely it must have been a flyer, or at least a glider?

After much study, and several failed attempts, a reconstruction was produced that actually flew. For reasons of cost, it was half life-size, and it had to be towed into the air with a stabilizing tail. But once airborne, the model flew unaided, without the tail. It glided on the thermals, using an onboard computer to make fine adjustments to the wings and head, and moving the counterweight projection behind its head to steer. The model provided proof that such an animal was a master of the air.

Careful anatomical drawings are made of the cleaned fossil bones. When done by a skilled artist, these drawings convey more information about the fossil than a photograph can. They are published along with a written description of a new fossil discovery.

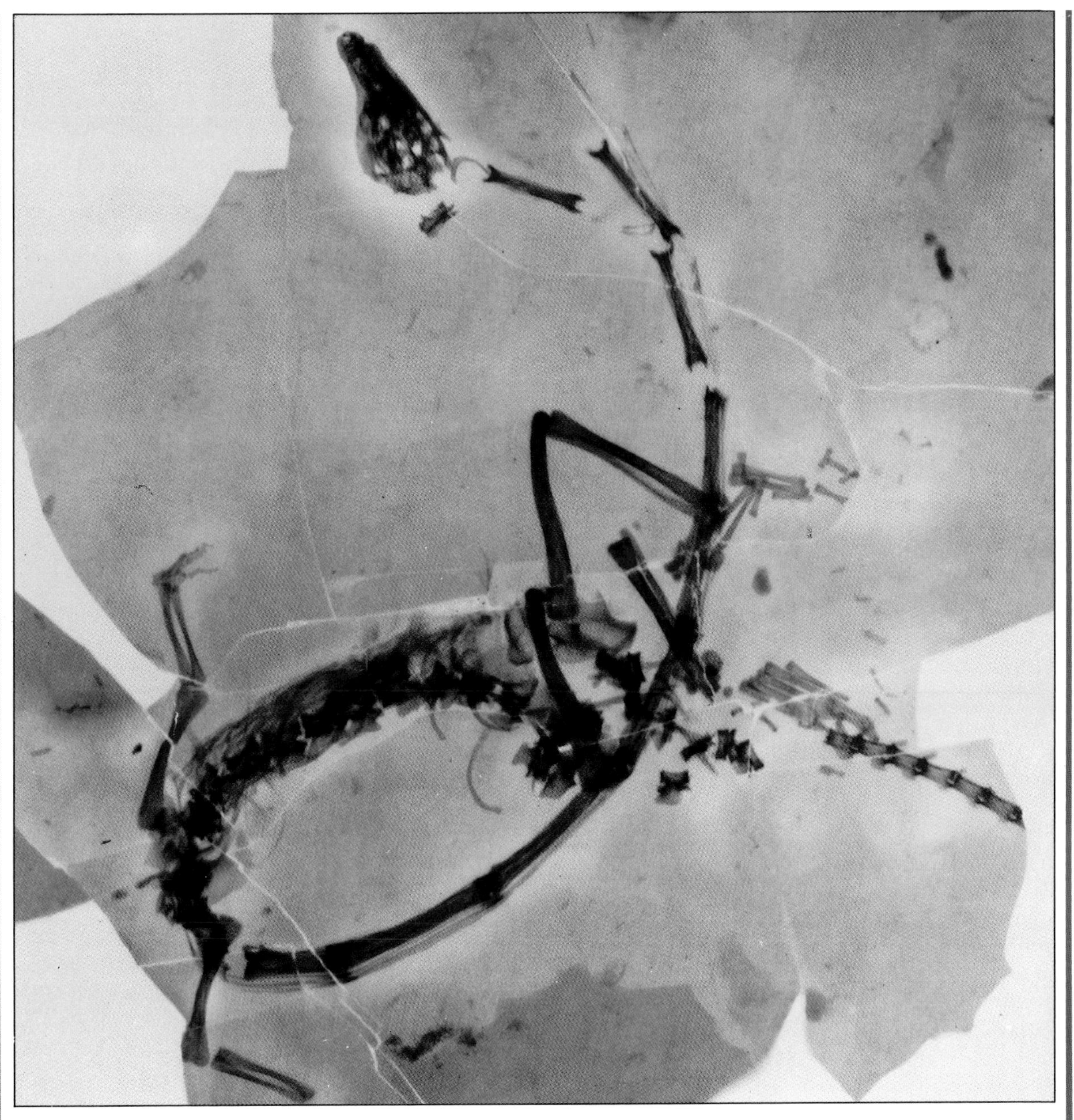

This fine X-ray photograph was taken through the Swiss mudstone which has protected these bones for 200 million years. This slim reptile, *Tanystropheus*, has nine neck vertebrae, each 12 in (30 cm) long. The extraordinary dimensions are a puzzle. The young of this animal were land-based insect-eaters, but the adults seem to have been marine fish-catchers. The amateur is unlikely to have access to full X-ray facilities, but such devices are often used by the professionals to check on the contents of a piece of rock before extraction begins.

Artists' impressions of prehistoric life are interpreted by model-makers for museums and "dinosaur" parks.

Above Life-sized models of early mammal-like reptiles – a carnivorous cynodont and an herbivorous dicynodont – as they might have faced each other about 210 million years ago. They are lurking in the trees of the Sudwala Dinosaur Park.

Right *Antrodemus* (*Allosaurus*), a fierce meat-eating dinosaur, prowls through a model desert land-scape at the National Dinosaur Monument in Utah.

Serial reconstruction

Series polishing has been used on primitive vertebrates and invertebrate shells. The fossil in its rock is polished away in small layers, and detailed drawings and photographs are taken at each level. These records are then transferred to layers of transparent material, which are laid one above the other, to build up the three-dimensional structure of the fossil.

A similar technique, employed for plant fossils, involves cutting in half the nodule bearing the specimen. It is then etched with weak hydrochloric acid, washed with water, and flooded with acetone. A sheet of acetate is smoothed onto the face and left to dry. When it is peeled off, it bears a cross-section impression of the water and sap bundles in the stem of the plant, which can be as detailed as a section cut from a living specimen.

Mounted skeletons

The impressive fossil skeletons of huge fish, early amphibians, giant reptiles and extinct mammals that line the galleries of major museums are the result of enormous effort. Hundreds of person-hours, specialized techniques and equipment, and in-depth expertise in all branches of paleontology, comparative anatomy, and related disciplines lie behind these displays. It is unrealistic for the amateur fossil hunter, as an individual, to expect to be able to reconstruct a dinosaur skeleton at home in the garage.

Nevertheless, opportunities do occur for the patient amateur to reassemble parts of a skeleton into more of a whole. If you have found several vertebrate bones, a selection of shell fragments, or the bark and leaf impressions of a tree, then you might wish to make a mounted display showing their relationships in life.

Work with plaster, latex or fiberglass casts of the fossils, rather than the original specimens. Accidents happen, and breakages occur, and a cracked cast is better than a ruined original. Casts are also lighter to handle and can be drilled for mounting rods and wires. Sketch out the relationships of the parts, and use brass rods or stainless steel piano wire for the framework – called the armature – on which they are mounted.

With some practice, it is possible to paint or stain a cast to a very close approximation of the original color. Try your artistic skills on unwanted bits of the casting material to see how they take the pigments. It may help to enlist the help of an artistic friend or colleague. After all, a professionally prepared reconstruction is likely to be the result of teamwork: from field collectors and laboratory specialists, to cast makers and professional artists.

Your fossils: a summary

The amateur fossil hunter is usually concerned with specimens other than giant dinosaurs or flying pterosaurs. Depending on the field of interest and the collection site, most specimens probably consist of shelled invertebrates, parts of plants, and a few fragments of vertebrate bones and teeth. Even so, keep anything you find that is potentially interesting: it may be significant to you or someone else in the future.

Reconstructing a broken or disarticulated fossil is like doing a three-dimensional jigsaw puzzle without the picture. However, with patience and developing skills, you should eventually identify and reconstruct your finds. You will be able to place them as plant or animal, and assign them to phyla; assigning a class may also be possible, but finding a genus may be more awkward. Identification to species level is difficult, even for experts.

If you have trouble and your fossil does not correspond to anything in the reference books, try taking it to a local museum, where there is usually someone to look at it. Most museums run such a service, although sometimes they charge a fee.

You will probably read descriptions about how the original organisms lived in the textbooks. But these are theories, and they are to be tested. You can try out the ideas on your own fossil. Do its features agree with the way of life suggested? You may be able to add to the body of conjecture by close study and clear thinking.

If you think your fossil is important and sheds new light on the past, a museum will confirm this. You may be asked to donate the fossil to their collection, and your own name may be used as a basis for the scientific name of a newly discovered fossil. If your specimen does turn out to be valuable, the fact that you received permission from the landowner of its discovery site will be all the more important. Your entitlement to the find may be contested if you did not obtain the correct permission.

TAKING IT FURTHER

Most people with an interest or pursuit benefit from meeting those with similar interests – to share their knowledge and tastes, learn from those with greater expertise and experience, and discuss the controversies of the day. Fossil hunting is no exception. You will also find out where to obtain equipment or how to improvise. (Major organizations and bodies concerned with paleontology are listed in the gazetteer section of this book).

Institutions

Major cities have museums and universities. Such institutions usually have a department of paleontology, or a related area such as geology. Smaller cities and towns also have museums, which tend to concentrate on particularly interesting aspects of regional history. If your area is famous for its fossils, they are sure to be represented here. There are also local geological societies that keep in touch with those people interested in rocks and fossils. The museum or library should be able to supply the name of a contact.

Museums that have paleontologists on their staff often offer an identification service for fossils. Or they may have literature to which you can have access, which should help in your hunt.

Museums and universities are a good port-of-call as you become interested in paleontology. Their collections of rocks and fossils may also be on display, to show you what is likely to occur in the locality.

Publications

Spectacular fossil finds, such as a new dinosaur, are often reported in local and national newspapers. But a steady stream of articles ("papers") about scientifically valid finds is published in the specialist periodicals and books often referred to as "the literature." They include general scientific journals such as *Nature*, *Science*, *New Scientist* and *Scientific American*, and more specialized journals, such as *Journal of Paleontology*, *Journal of Vertebrate Paleontology*, *Paleontology* and *Paleobiology*. A good library or book store can order them for you, or you can write to the publisher for details. "The literature" keeps readers in touch with recent methods and finds, and carries advertisements for manufacturers, retail outlets, and mail-order companies that sell equipment for the fossil hunter.

The specialist journals tend to use technical (although precise) terms, and they can be hard for a beginner. So it is wise to start with more general publications and progress as you "learn the language." Also, do not expect the location of an exciting new site to be freely advertised. Locations are sometimes kept from the glare of publicity until the scientific work is completed, in case of attention from unscrupulous fossil dealers or careless trophy hunters.

Museums are usually the ultimate destinations of fossils, especially valuable ones. They are the centers of local or worldwide expertise in paleontology. Here, also, is the expertise to display fossils in an informative manner. Visits to national museums at home or abroad are special treats for most enthusiasts. It may not be possible to deal with all the fossils on display in one visit, so be selective for maximum benefit. There may be facilities for viewing fossils that particularly interest you, but which are not in the main display. Always ask.

SEPTEMBER 1988
$2.50
U.K. £1.50

SCIENTIFIC AMERICAN

A new kind of cancer gene derails normal cell growth.

How the body makes insulin—new insights, new clinical promise.

Laser spectroscopy—powerful key to atomic discovery.

The Fossils of Montceau-les-Mines

Some 300 million years ago central France lay at the Equator. The paleoecology of this bygone world has been reconstructed from a superb fossil cache

by Daniel Heyler and Cecile M. Poplin

The final part of the Paleozoic era, some 300 million years ago, was a time of transition. The Carboniferous period, when the continents were gathered in several landmasses near the Equator and a hot, humid climate sustained the swamp forests that gave rise to the major coal reserves of today, was drawing to a close. The world was about to enter the Permian period, when humid heat would give way to a cooler, more arid climate and the first reptilian ancestors of mammals would begin to populate the continents.

Central France was then a region of hills covered with giant ferns and conifers, interspersed among lowlands patterned with rivers, lakes and lagoons. Millipedes, scorpions, insects, salamanderlike amphibians and reptiles crawled on the land, and the waterways teemed with worms, crustaceans, mollusks and ancestral sharks and fishes. When these creatures died, their bodies sank to the bottom of swamps, where they were preserved and eventually fossilized.

Some of these fossil plants and animals were unearthed during the 19th century at Montceau-les-Mines, a coal basin situated northeast of the Massif Central, a mountain range formed during the Paleozoic at about the same time as the Appalachians. It was only about 10 years ago, however, that strip-mining of the Montceau coal seams revealed the magnitude of the fossil deposit. Many paleontologists now agree that it is one of the richest discoveries of the past decade. Galvanized by the imminent destruction of the site, local amateur paleontologists converged on the mine. Weekend after weekend they raced to save fossil material before it would be destroyed by strip-mining machinery.

Working with the Paleontological Institute at the Museum of Natural History in Paris and the Natural History Museum in Autun, near Montceau, they harvested 7,000 shale slabs and more than 100,000 nodules, most of them bearing fossils. They carefully recorded the location and orientation in which each nodule was found. Later, when the fossils had been identified, it would be possible to reconstruct their distribution in the fossil beds. This was the first time such precise, modern excavating methods had been employed on such ancient deposits.

The mine, no longer operational, has been filled in. For the scientists involved, the enormous task of examining the rescued specimens is still under way. About a fourth of the nodules have been opened, and nearly 22 percent of them have yielded well-preserved fossil animals. Many specimens proved to be in excellent condition; they included such animals as insect nymphs and soft-bodied worms, rarely found in a fossilized state. A research effort coordinated by us and supported by the National Center for Scientific Research (CNRS) ... possible to assem... team of specialis...

first lot of fossils. The workers have been able to reconstruct the main features of the site's plant and animal community as well as of its environment. To date nearly 300 species of plants and pollen and 16 classes of animals representing about 30 genera have been identified.

The fossils represent a slice of the earth's geological past that stretches from the end of the Carboniferous to the beginning of the Permian. During that time the region's tropical climate gave way to a more temperate one, accompanied by changes in the flora and fauna that made them better adapted to the drier climate. Thus the deposition documents the transition of species over a critical period in earth history.

The fossils of the Montceau basin are either shale deposits, formed when animals and plants were flattened between layers of silt, or nodules. The process by which the latter came about is not completely understood, but it is thought that an organism can act as a nucleus around which fine sediments accumulate to form the nodule. Nodule fossils—also found at contemporaneous sites such as Mazon Creek near Chicago and at later ones such as the Triassic site in Madagascar and the Cretaceous site at Ceará in Brazil—are more three-dimensional and so give a clearer picture of an organism's physical form.

The zoological groups represented in the Montceau fossils vary considerably in the number of individuals. For several reasons, however, the numbers do not accurately reflect the faunal populations in the ...

DANIEL HEYLER and CECILE M. POPLIN are paleontologists at the Museum of Natural History in Paris (Unité Associée 12 of the National Center for Scientific Research). They share an interest in Upper Paleozoic vertebrates. Heyler specializes in fossil fishes, amphibians and tetrapod footprints from continental basins of western Europe, in particular France. Poplin studies mainly North American fossil fishes. She has also worked on European and Chinese fishes and primitive mammals from Mongolia. Montceau researchers not mentioned in the article include J. Doubinger, G. Gand, Jean R. Langiaux and M. D. Sotty.

70 • SCIENTIFIC AMERICAN September 1988

mining made it impossible to collect all the fossils. Last, the mine is only a part of the entire geological basin. Nevertheless, for sheer quantity, quality and diversity of specimens, Montceau can be compared to the famous, contemporaneous site at Mazon Creek and the Bear Gulch, Mont., site (where the fauna is about 15 million years older).

The oldest layers at Montceau were deposited at the end of the Carboniferous period, which began 345 million years ago and lasted for about 65 million years. The period accounts for the world's coal reserves. Coal formation slowed as the climate became cooler and drier during the succeeding Permian period, which began 280 million years ago and lasted for 50 million years. Hence coal from the Permian is scarce. The transition between the Carboniferous and the Permian periods is not marked by an abrupt geological change and so is not recorded as a stratigraphic boundary. One must turn to paleontology and in particular to paleobotany in order to assign a deposit to the transition.

It was the fossil flora more than the fauna that enabled the lower layers at Montceau to be dated to the Stephanian (the last stage of the Carboniferous) and the upper layers to the Autunian (the first stage of the Permian). The evolution of the flora throughout Montceau bears witness to the change in climate that took place between these two geological periods.

ANIMALS that lived during the Upper Carboniferous period, 300 million years ago, are preserved in these fossils found at Montceau-les-Mines. ...

New Scientist 5 August 1989 — 31

SCIENCE

Flying dinosaur flapped its wings

Sue Bowler

PTEROSAURS did not simply glide. They were active fliers and grew hot as they flew. The result comes from examining the best preserved fossils of pterosaur wings that have ever been found. David Martill of the Open University, Milton Keynes, and David Unwin of the University of Reading discovered tissue from a pterosaur's wing in Brazil. The tissue has more in common with bat wings than with the skin of modern reptiles (*Nature*, vol 340, p 138).

The pterosaur, known as *Sandactylus*, lived about 100 million years ago. It probably had a wingspan of about 5 metres.

The surface of the pterosaur's skin looks just like the skin between a human thumb and first finger. None of the original organic material remains. Nevertheless, Martill and Unwin could see fine structure of the thin top layer of skin, known as the epidermis, and the thicker layer below, known as the dermis. With the aid of a scanning electron microscope, the researchers could even see the muscle fibres beneath the skin.

The results show that the wings of this pterosaur were not just membrane. Within the upper layer of the dermis, Martill and Unwin found structures that they identified as blood vessels. Bats have similar arrangements of blood vessels, which cool them down as they fly.

If pterosaurs needed to cool down, say the palaeontologists, then they must have worked as they flew. In short, they must have flapped their wings, rather than gliding.

However, Martill and Unwin found no signs of fibres within the wing. These would be needed to stiffen the wing as the pterosaur flew. The researchers suggest that *Sandactylus* kept its wings at the correct tension by moving its back legs. These were attached to the wings. This means that pterosaurs had more control over their flight than scientists had previously thought. They were not simply passive gliders.

Previously, Soviet scientists described a pterosaur wing from an animal that was a distant relative of *Sandactylus*. Their findings, which included fur, were quite different, however, from those reported by Martill and Unwin. □

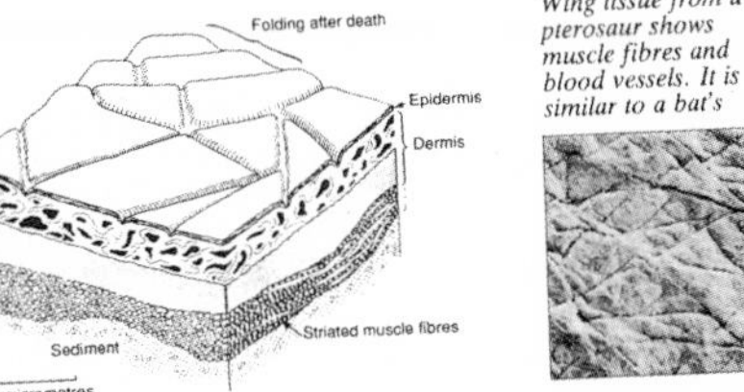

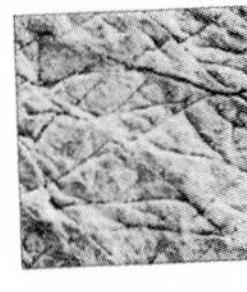

Wing tissue from a pterosaur shows muscle fibres and blood vessels. It is similar to a bat's

The popular and scientific press often carries articles on the latest impressive finds or on current arguments in paleontology. The universal interest in dinosaurs and anything prehistoric guarantees the value of such features. If the subject of the article is of particular interest to you, write to the person or institution involved, or to the publication's editor.

JOINING AN ORGANIZED DIG

The universities and large museum departments are the main organizers of digs at significant fossil sites. These digs serve to improve their collections and provide new material for study. Such expeditions are very expensive to run and need to be planned well in advance. The professionals involved work mainly in advisory and supervisory roles, once the dig is underway. Many other pairs of hands are needed at a large site, just for the purpose of moving rock and earth, as well as for the routine work of mapping, recording, and fossil extraction. Students from the university are usually present, but the rest of the work force may be recruited through advertisements at the local museum.

Spending time on a professional dig is invaluable – it is one of the best ways of learning about practical paleontology. If you can spare some time, watch for announcements in the relevant publications (those specializing in paleontology are most likely). The job will consist largely of manual labor, and you will work as a volunteer.

The work and the rewards

The kinds of jobs available on a site include digging and chiseling out rocks, removing them from the site, sieving and sorting tiny fragments that may contain the occasional tooth, and scraping and brushing away soft rocks above a fossil bone. You may spend hours in uncomfortable positions, working outside where it always seems to be too hot, too cold or too wet. You will probably get covered with dirt, and have to help carry large rocks over difficult terrain. There may be no modern sanitary facilities near the site, and part of the work regimen could involve keeping others supplied with food and drink. A big find will take years to clean in the laboratory, and it might be a decade or more before you can stand back and inspect the fruits of your labor.

But there are many compensations. These can include new friends, shared interests and expertise, practical "hands-on" experience, and incidental tuition as experts visit your part of the dig for an examination and discussion. There is also a tremendous sense of achievement when a new specimen is revealed, or when a fossil at last comes free and is ready to be moved to the laboratory.

Above On an organized dig, you will be expected to work – but you will also spend time with like-minded people. Expensive equipment is provided, and the chance of success will be greater when working with experts. Here, the team is planning how to unearth a dinosaur.

Right A professionally organized field trip to a known fossil site is another way of gaining experience in the practical side of paleontology. You are more likely to find something interesting, and experts are available to help you interpret your find.

FINDING FOSSILS IN SURPRISING PLACES

Fossil hunting does not always involve expeditions to exotic beaches or remote cliff faces. Fossil-bearing rocks are all around, and you may be able to collect or study closer to home than you think.

In your own backyard

Your backyard or park, no matter how small, may contain rocks: there might be a gravel path, or a rock garden with alpine plants, or merely irritating boulders in the flowerbed. A thorough search might produce the basis for a fossil collection.

The gravel bought from garden centers often contains fossils, depending on where it was quarried. Finds include shells of small ammonites, belemnites, bivalves or the brachiopod *Gryphaea*, and ossicles or fragments of crinoid arms. Examine the pebbles for regular shapes and patterns or for traces of different textures on the surface. Take a hammer and break some open, just in case.

Digging in the backyard may reveal fossils. Many agricultural areas consist of drift sediments laid down by receding glaciers. The soil may be an unsorted mix of fine clay and larger lumps; flint boulders are often found in clay soil which overlies chalk. Although the boulders often bear a resemblance to large vertebrate bones, the fossils they contain are likely to be seabed invertebrates such as corals, sponges, mollusks, barnacles and echinoderms. Some of the boulders may be ironstones, which could contain plant fossils from the great coal seams.

Your space may also unearth such treasures as mollusk shells or vertebrate bones and teeth. Bones in the yard may well be the remains of a meal that is fairly recent (in geological terms), or of the animals that helped people to live and farm in the days before mechanization. They are not fossils in the strict sense. But the only way to be sure is to identify them correctly.

If you use coal, split a few pieces along the bedding plane. You may see the frond shapes of the ferns that once grew in the coal forests.

Built-in fossils

If your local buildings are made of stone, a look at the walls may prove fruitful. Many buildings are made of igneous rocks, such as granite, while newer structures are often of cement. However, some older buildings were made of sandstone, which is relatively tough, yet easy to work. You can identify sandstone by the rounded particles of sand and the layered appearance; the characteristics of the layers show how the sediments were laid down.

"Flagstone" is a sandstone which is often used to make patios. It was formed at the bottom of a flat river bed, by water flowing either very fast or very slowly. Churches and other old buildings often show the wavy patterns in sandstone produced by slower water currents. You may also see the layers from desert sand dunes or a bed of pebbles from the shore.

Limestone such as Portland stone may be the remains of tropical coral islands, and is mostly made up of the skeletons of corals, shellfish and

Cylindroteuthis, a Jurassic belemnite. These fossils are the internal shells of cephalopod mollusks, found in sediments from beneath Mesozoic seas.

Lepidodendron, a giant clubmoss from Carboniferous forests. This one was discovered in a coal field.

protozoans. The fossils in some limestones can also tell you how the rock was formed: the more complete the remains, the gentler the currents, while fragmented fossils indicate rougher seas. Portland stone may be full of shell holes, when it is known as "Roach."

You cannot collect these fossils, but with permission you can take close-up photographs and perhaps a mold of modeling clay for your collection. The slates found on old roofs are metamorphosed shales, and occasionally they contain fine fossils, such as graptolites or small fishes. Look for them on demolition sites.

The Fossil Grove in Glasgow, Scotland. Five stumps of scale trees, 350 million years old, were discovered when an abandoned dolerite quarry was made into a park in 1887. These trees were growing at the same time as the forests which became the Scottish coal seams.

SPECIALIST INTERESTS – NARROW YOUR FIELD?

The fossil record, like the living world, is vast and bewildering. All the groups of animals and plants alive today are represented, together with many that flourished, often for many millions of years, but then died out. As you begin to find and study fossils, you will probably soon discover that it is not possible to be knowledgeable about anything more than a small segment of the record. It may be more satisfying to concentrate on an area which few others cover. Paleontology covers so many organisms, over such long timespans, that this is relatively easy to do.

The factors governing your study area – travel, finance, time, other commitments, space and so on, have already been discussed (see page 38). But one of the fascinations about fossil hunting is that the more you look, the more you see. As you gather more information, so the wonder and satisfaction increase. Give yourself time. If the fossils in your locality do not at first appear especially spectacular, they may do so as you find out more.

Collecting considerations

Building up a large collection is a prime aim for some fossil hunters. But collecting for collecting's sake is no longer a satisfactory reason for damaging the environment. Try to direct your activities toward a definite line of study. Begin by looking at the fossils already in your collection, or in that of a colleague. One particular group may keep popping up, which is common in your area. Begin here and see if you can find some pattern in these fossils. Do they change slightly with time? Is there much variation among contemporaries? Form some sort of simple theory about differences and similarities, and set out to apply it to new finds or other collections in the vicinity. Proving, disproving or enlarging upon your hypothesis by further collecting may soon take priority over collecting for its own sake. The line of study has given direction, significance, and selectivity to your future searching.

This type of sequence is the basis for scientific progress, and it will add meaning to your collection. It can also help by curtailing the equipment you need, such as tools and books, and therefore cutting down on your expenses.

Phyllograptus graptolites from Ordovician rocks. Graptolites survived from the Cambrian to the Carboniferous periods, but they are most important for stratigraphical identification of Ordovician and Silurian rocks. Looking like the mythical "writing on the wall," these fossils are the skeletons secreted by colonies of worm-like animals. They lived in the sea: some settled on the bottom, others floated at the surface. Their colonies grew in many different patterns and are a subject for specialization. They are found all over the world.

Asteroceras obtusum, the internal mold of a later ammonite from Upper Jurassic rocks. The features used to distinguish between different ammonites are the tubercles, ribs and ridges or the shell, and the delicate tracery of the suture lines between the shell compartments.

Gastrioceras listeri, a goniatite from Carboniferous coal measures. The goniatites were early members of the Ammonoidea group, to which the ammonites themselves belonged. The various kinds of ammonoides form stratigraphic indicators, especially in Mesozoic rocks.

SPECIALIZATIONS WITHIN PALEONTOLOGY

As with many branches of science, it is difficult to separate paleontology from its parental sciences, and from its many sibling disciplines. It occupies a niche between biology, the study of the living world, and geology, the study of rocks. Paleontologists need to be familiar with the basics of both these scientific neighbors. There are many areas within paleontology, and many subjects closely allied to it, that can exert their attraction on the fossil hunter.

Biostratigraphy

A science begun by William Smith (page 24), biostratigraphy is based on the Principle of Superposition and the Law of Strata Identification by Fossils – strata of the same type and age contain the same fossils, given similar conditions. The rocks are dated using such information.

Rocks are classified into the names of the eras or periods during which they were formed. Each system has a stratotype, a sequence of rocks and fossils to which other rocks can be compared. This is analogous to the classification of plants and animals, where there is a *type specimen* for each species, to which others are compared. The *type stratum* is usually the one where the system is first recognized.

Systematic paleontology

Although paleontology is the study of "ancient life," in practice, it is often taken to cover the study of animals. Paleontologists may be interested in one main type of animal and its history during a certain time span. For example, paleoichthyologists study the history of fishes, and paleoconchologists follow the evolution of mollusks. Some students are interested in vertebrate or invertebrate animals during the various stages of their development and life cycle. Others study the puzzles of taxonomy, the classification of species. Evolutionists look to the fossil record when examining how one species might change into another.

Paleobotany

The study of the geological history of plants is termed paleobotany. The first person to treat fossil plants systematically was Adolphe Brongniart (1801-76). He analyzed fossils from all over the world, usually in the form of carbonized remains from coal seams, and he later included permineralized fossils in his studies. He also realized that such remains could be used to work out the age of rocks.

Life began in Precambrian times, in the form of simple plant-like organisms. Studies of the way these simple blue-green algal cells evolved is shedding light on the crucial stages in the development of living things: how life itself appeared; how cells without nuclei (prokaryotes) evolved into more complex plant cells containing nuclei and other organelles (eukaryotes); how sexual reproduction began; how single-celled organisms became multi-cellular organisms; and how aquatic plants were able to change and colonize the land, to be followed by animals. The ever-increasing discoveries of early plant remains in very ancient rocks is providing a stratification system for the relative dating of Precambrian rocks.

The study of ancient plants itself has several branches. Examining and interpreting microscopic plant fossils, particularly pollen and spores, is known as palynology. Pollen grains are very resistant to degradation and are often identifiable under a high-power microscope, from the characteristic sculptured patterns on their surfaces. Pollen is only produced by land plants, but because grains are released in their billions and transported by the wind, sediments from bodies of water usually contain them. Palynologists count the relative numbers of grains and quantify the types of plants in past land environments. The conditions required for fossilization mean that relatively few land organisms have left a record; pollen is the best "window" we have to the past. Pollen grains are also very useful as stratigraphic markers, because they are so widespread in the fossil record. Pollen stratigraphy is yet another branch of paleontology, important to those looking for fossil fuels.

Trees have been preserved in amazing detail as their tissues were permineralized by crystals of opal. The molecules have been replaced one by one, and so the fine details of cell structure are revealed when thin slices of fossilized tree trunks are viewed under high-power microscopes.

The science of dendrochronology, the study of tree rings, is used primarily to date ancient woods in archaeology. It relies on the fact that most trees

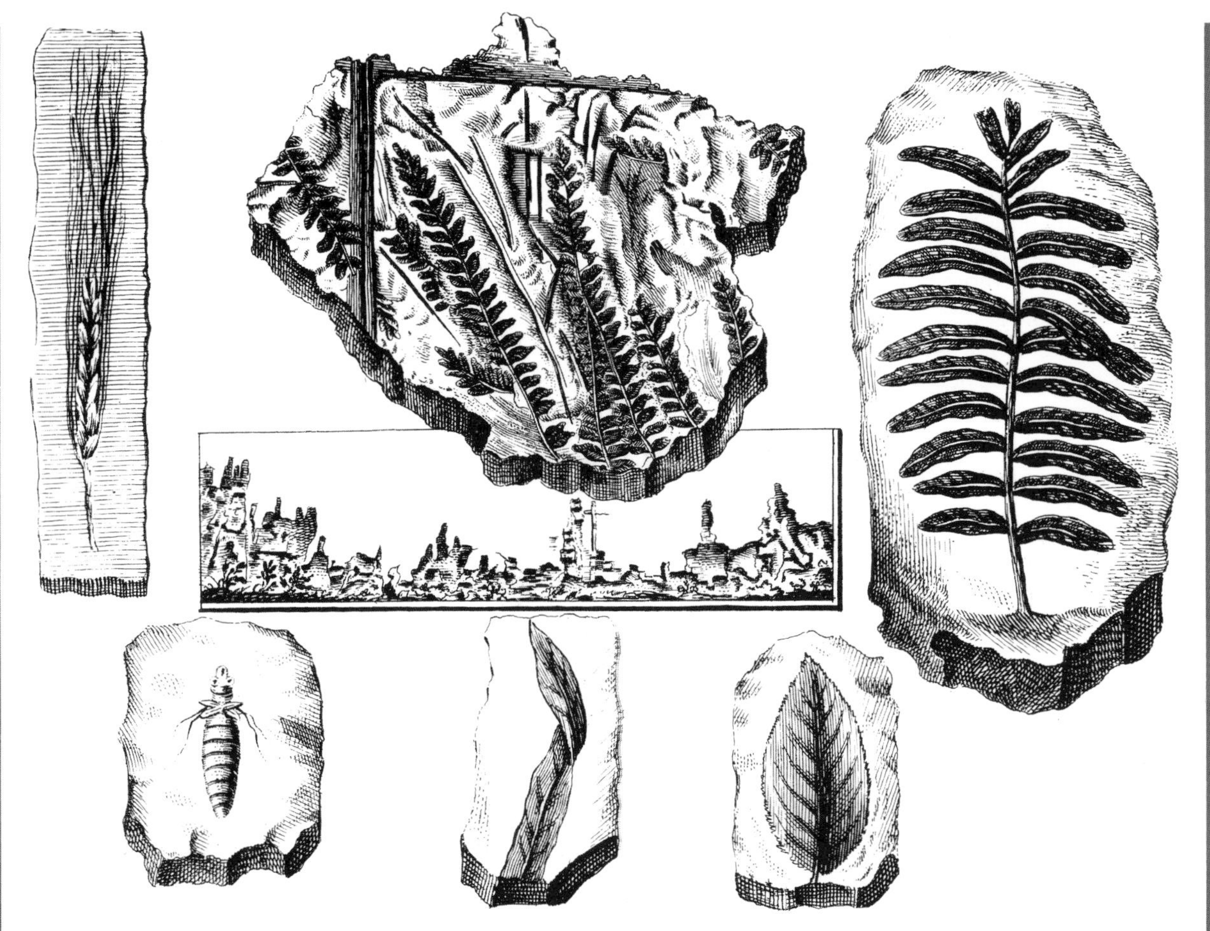

grow by producing a ring of new tissue each year. The thickness of the ring depends on growing conditions: a relatively warm, wet year produces a thick ring, while little new wood is laid down during a dry or cold year. These patterns can be followed and correlated between trees and woodlands. Using very long-lived trees, such as the bristlecone pine, an unbroken record has been established going back 8,000 years. Although the fossil record is far older and sketchier, the science can be applied to fossilized tree trunks, to analyze the climate of the time.

Micropaleontology

In each sample of sedimentary rock that bears fossils, there are huge numbers of tiny organisms that can only be seen under a microscope. This branch of the science is known as micropaleontology. The fossils are mainly of single-celled organisms such as foraminiferans, radio-

Illustrations from *Nature Displayed*, published in 1740. This work was a translation of *Le Spectacle de la Nature* by La Pluche. The impressions of ferns, leaves and grass, called "figured stones," are beautifully drawn, even though the fossilized plants of paleobotany were not as popular as the finds of fossil animals at the time.

larians and diatoms. Some very small multi-cellular organisms can also be seen. The electron microscope enables scientists to study fossils as small as 0.005 mm across. Such work is applied to the detection of new oilfields. It is also invaluable in helping paleontologists to look for the earliest forms of life in the rocks.

Paleoecology

Paleoecologists are interested in how prehistoric animals and plants lived together in their habitats, and how they filled the available ecological niches in the environment. Where the fossil record is rich, it is possible to calculate the relative abundances of plants, herbivorous animals, and carnivorous animals, and so to propose a kind of food web. Recent, very careful studies have begun to suggest the ethology, or behavior, of animals that lived in the past. It is possible to suggest how they obtained and ate their food, avoided predators, courted mates and reared their offspring.

Paleoclimatology

The kinds of organisms that existed in the past also suggest the types of climate in which they lived. This branch of the science is paleoclimatology. Temperatures, rainfall, wind patterns and ocean currents can be estimated. Fossils of sea creatures show the changing levels of the oceans. These in turn indicate how much water was bound up as ice, which is a pointer to global temperatures. Evidence demonstrates, for example, that the coal forests of the Carboniferous period must have flourished in a warm, moist climate. Dendrochronology and palynology also provide valuable clues to climates in ancient times. The fossil evidence is matched with the sorts of sediments laid down, such as unsorted tillate from glaciers, or well-sorted river deposits.

Paleobiogeography

Many areas of paleontology are used by people who study the physical history of the Earth. Whereas biostratigraphy looks at the rock record vertically, so paleobiogeography looks at it horizontally across the world at a particular time level. The places where fossils are found yield a great deal of information about the geographical distribution of plants and animals in the past. Indeed, a breakdown of the word paleobiogeography shows it means the study of the geographical distribution of ancient life forms.

This branch is divided in turn into paleozoogeography, for ancient animals, and paleophytogeography for plants. Among its other achievements, this area of paleontology has helped to chart the wanderings of continents across the Earth's surface, as explained by the modern theories of continental drift (page 26). Similar fossils from similar strata in eastern South America and West Africa are a strong indication that, during that time span, these two land masses were joined or very close together. This leads to the field of paleogeography.

Corals are a fascinating subject for study and a good example of a systematic specialization. Corals were among the first animals with skeletons, and have been building reefs in shallow seas ever since. This is a Silurian coral, *Acervularia ananus*. It shows how the colony is made up of individual conical corallites with radial septa (separating "walls").

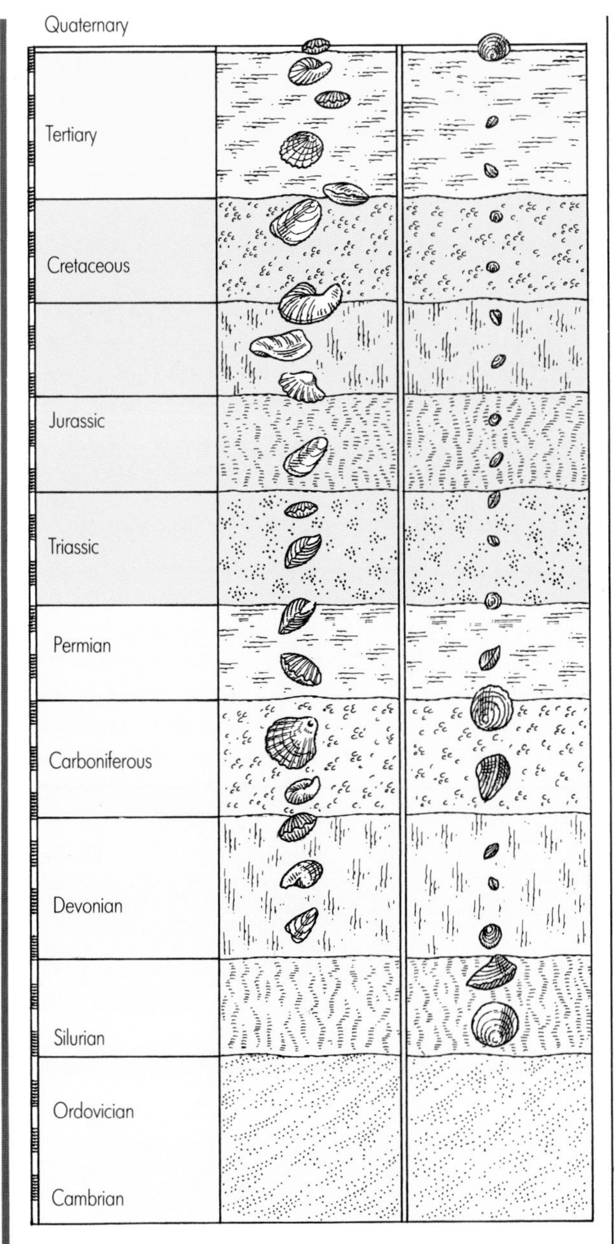

Above The gradual evolution of common fossils over time is one suitable field of specialization. The brachiopods (lampshells), shelled animals outwardly resembling mollusks, have a long history. Some have survived to the present day while others went through rapid periods of evolution.

Writing the history of the Earth

Strangely shaped pieces of stone – fossils – have helped us to build a picture of the Earth's history from near its beginnings. The physical world has constantly changed through time, and the cycles of crust movements, rock erosion, and climatic changes continue today. The history of living things is closely interlinked with the physical world, and we now know much about its evolution.

Yet the fossil record leaves many unanswered questions. The stratigraphic systems are marked by great changes in the plants and animals preserved, with sudden mass extinctions or "explosive evolution" when many new forms quickly appear. We still have little idea why many such events occurred. These are the challenges for fossil hunters in the future.

The Earth in the Triassic

The Earth today

Lystrosaurus

Above The theory of continental drift is based on the fact that similar rock types containing similar fossils are found on continents now widely separated. About 250 million years ago, most of the land masses were linked together in the giant supercontinent called Pangaea. Since that time, they have crept apart to their present positions. *Lystrosaurus* was a mammal-like reptile that led a hippopotamus-like existence during Triassic times. Its fossils are found in Africa, Antarctica and India, indicating that there were once links between these lands.

CONTROVERSIES IN PALEONTOLOGY

The world of fossils is never static. Paleontology, like other sciences, is continually growing and evolving, and has its fair share of heated discussions, controversies and colorful personalities. Are the microscopic flecks found in early rocks really the first traces of life? Who were the first true humans, and where did they come from?

Two examples, one past and one present, serve to illustrate how our ideas about fossils must continually be updated and re-interpreted in the light of new evidence:

The Piltdown hoax

The search for human ancestors began to gather momentum toward the end of the nineteenth century. Remains of Neanderthal Man (1856), Cro-magnon Man (1868) and Java Man (1890s) had already been discovered with increasing degrees of interest – and various interpretations. Human-like skull fragments were discovered in a gravel pit in the south of England in 1912 and seized upon by the very receptive paleoanthropologists of the time. The skull was large, and a matching ape-like jaw was soon found at the same site. Named "Piltdown Man" after its discovery location in Kent, the new fossil appeared to be the missing link between man and apes that everyone was so eagerly anticipating.

Much argument, research and publication took place concerning these finds. Yet it was not until 40 years later that fluorine dating techniques revealed the "fossil" as a fake. Bones that have been buried for similar lengths of time absorb similar levels of the chemical fluorine; yet in this case the skull and jaw did not contain the same levels of fluorine. The skull was of a relatively modern man, and the jaw was of an orang-utan with teeth glued into it. No one knows who made the fake. But it is interesting that some scientists of the time saw what they wanted to see, rather than what was actually there.

Even in the early part of this century, many people found it difficult to consider that human beings had evolved anywhere except Europe. The early European hominid finds had reinforced this view. Consequently, when Piltdown Man appeared in a very "civilized" part of the world, it was exactly what scientists of the time wanted.

Meanwhile, in South Africa, Raymond Dart had found the remains of a skull which was neither ape nor human. This discovery was treated with scorn by anthropologists. Many held the view that South Africa was not a suitable place to be the "Cradle of Mankind." Again, prejudice and expectation interfered with good scientific analysis. It is now believed that much of early human evolution did indeed take place in Africa.

Since the exposure of the Piltdown fake, all significant fossils are routinely subjected to rigorous testing.

Dinosaur discussion

The second example concerns the many discussions surrounding that most popular of fossil groups, the dinosaurs. The origins of dinosaurs in

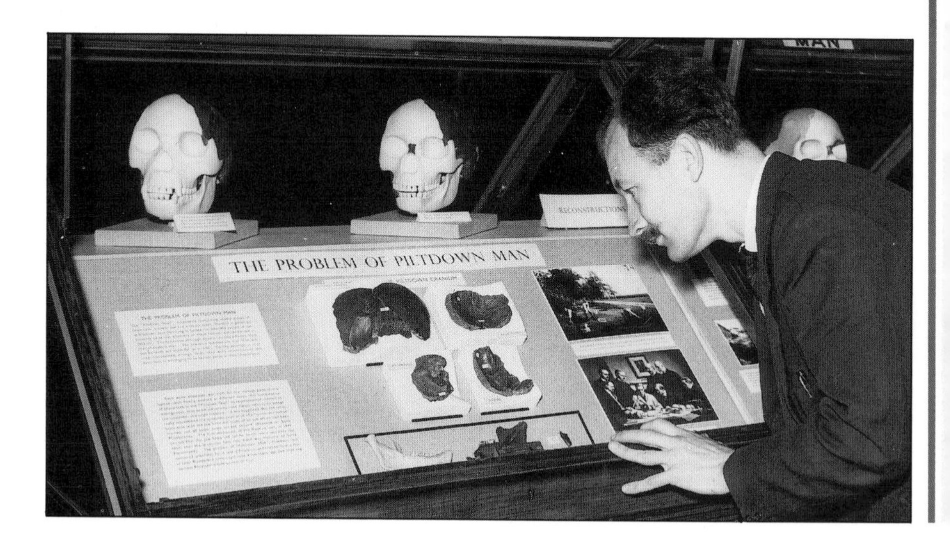

In 1953, the British Museum (Natural History) had to rearrange its fossil displays to take account of the newly-discovered fact that Piltdown Man was a fake. Who carried out the hoax is yet another detective story for paleontologists.

the early Triassic period, and why mammals did not begin their dominance at that point, is one frequently aired question. Was it that the dinosaurs had developed a variable gait using their semi-erect legs? Was the arid climate of the time suited to them? Or did a mass extinction of the animals that had gone before leave niches most suitably filled by the dinosaurs?

Were the dinosaurs cold-blooded or warm-blooded? The factors that some believe point to the latter interpretation are as follows. Many dinosaurs were bipedal, as are birds and some mammals; the raised head posture of many dinosaurs meant that they must have had a powerful and fully divided heart, like birds and mammals, but unlike modern reptiles; although most dinosaurs had small brains, some had larger-than-average reptile brains, another warm-blooded characteristic; some dinosaurs lived in apparently cold climates; and predator-prey ratios of fossil remains are similar to those of modern mammal predator-prey ratios. The discussion surrounding these theories and the study involved has done much for the progress of dinosaur fossil interpretation.

Even over a single animal, debates have flourished. The impressive plates that lie along the back of the dinosaur *Stegosaurus* have been the subject of argument. Were they "armor-plating" for protection of these herbivores, or "solar panels" to facilitate warming by the sun?

The ultimate dinosaur controversy is the famous "Death of the Dinosaurs" debate. Various theories for the mass extinctions at the end of the Mesozoic period have been put forward. The most popular is the death of plants, and therefore of all life that depended on them, caused by the dust cloud after a meteorite collided with the Earth.

Another controversy has surrounded the fossils of the earliest bird, *Archaeopteryx*, from Germany. There has been continuing debate about whether it is a reptile with feathers, or a bird with reptilian features. Lately, there has even been discussion as to whether the fossils are real or fakes. One theory put forward is that the reptile-bird "missing link" fossil is actually a *Compsognathus* (a small dinosaur) fossil overlaid by a thin layer of sandstone cement, into which feathers had been carefully spread by human hands, to leave their impressions. Extensive research by the museums that hold *Archaeopteryx* fossils seems to have proved that the fossil is indeed real, and laid to rest this argument.

Stegosaurus means "roofed reptile." When it was first discovered, the bony plates arranged in opposite pairs and lying flat across its back were thought to form a protective "roof." More discoveries indicate that in fact the plates probably stood erect, in two staggered rows along the back. The dinosaur would probably have stood sideways to the sun in the morning, in order to warm up. In the heat, it would seek shade or stand head-on to the sun, letting the breeze cool the plates.

THE EVOLUTION OF LIFE

In these next four sections we look at the appearance, development and change in living things over the vastness of geological time. To understand how life began and progressed from one form to another, it is necessary to understand a few basic biological principles.

Living things are complex arrangements of molecules that hold themselves together in a structured way, absorb nutrients and use energy, sustain themselves, grow and reproduce. Life began as the simplest arrangements of chemicals. Eventually the collections of molecules became simple cells, and they in turn became more complex cells. Somehow, a number of single-cell units stayed together and began to rely on each other, each performing different tasks for the benefit of the colony. The multi-celled organism appeared.

Life in the laboratory?

We may never know how life began; there are no records. The first complex molecules that may have hung together in the primeval seas are not in the fossil record. In the 1950s, U.S. scientist Stanley L. Miller devised an experiment in which he simulated the primeval sea and the beginnings of life. He made a solution of the chemicals likely to have been in the ocean at that time, and introduced an atmosphere of ammonia, methane and hydrogen above it. He then subjected it to electrical discharge – the "lightning" of an electric arc. Miller heated this solution for a week and found that various complex organic molecules had formed.

These experiments have since been refined, and many of the complex molecules found in living tissue have been created. But science has not yet managed to create a collection of molecules that is able to sustain and reproduce itself.

The guiding force

Charles Darwin proposed that the force behind evolution was natural selection. As we see today, no two organisms are exactly alike. The processes of reproduction mean that organisms usually inherit a mixture, a mosaic or a combination of characteristics from their parents. Some individuals are therefore better equipped, or adapted, to survive in the environment in which they live than others are. In general, the longer individuals survive, the more offspring they will produce, most of them carrying the characteristics that benefited their parents. Those that inherit characteristics which make them less able to survive will be less likely to produce offspring, and therefore those characteristics will be lost.

These inherited characteristics are governed by the genes, groups of chemicals on the chromosomes found within the central nucleus of each cell. The science of genetics examines how the genes are passed on from parents during sexual reproduction, and how they are expressed to control the physical and chemical features of the individual.

A theoretical framework

Put simply, evolution is based on the principle of "survival of the fittest." The process is very slow, occurring over hundreds or thousands of generations. As a theory, evolution is almost impossible to test, and there have been many suggested refinements. For example, some evolutionists have suggested that the changes we see between species happen suddenly, followed by a period of stability. Others consider that the process continues at a gradual rate. But as a broad framework, the theory of evolution has stood the test of time, and it provides a basis for studying the changes we see in the fossil record. Evolution must have steered the developments of living things since the very beginnings of life.

The following four pages contain two reference charts. The first shows the periods of geological time, with a summary of climatic conditions and the main features of plant and animal life for each. The second chart illustrates how animals and plants have evolved into many different groups, providing a classificatory framework into which we can incorporate fossil finds.

190 million years of evolution separate the owner of these feet and the creature that made the fossilized footprints, the reptile *Cheirotherium*, in Arizona. Evolution has "invented" structures such as feet, wings, teeth and shells several times.

GEOLOGICAL TIME CHART

	PALEOZOIC ERA – 345 mill. yrs.						MESOZOIC ERA –	
	CAMBRIAN PERIOD From 570 million years ago	ORDOVICIAN PERIOD From 500 million years ago	SILURIAN PERIOD From 430 million years ago	DEVONIAN PERIOD From 395 million years ago	CARBONI-FEROUS PERIOD From 345 million years ago	PERMIAN PERIOD From 280 million years ago	TRIASSIC PERIOD From 225 million years ago	JURASSIC PERIOD From 193 m years ago
GEOGRAPHY AND CLIMATE	• North America and Eurasia separate • Gondwana-land lies in the south • Sea levels rise, sandstone and limestone deposited • Atmospheric levels of oxygen rise	• All land masses move south • Extensive glaciation in areas which now lie near the equator • Floods subside but return and cover more land than at any other time	• North America and Eurasia join, producing Caledonian Mountains • Gondwana-land covers South Polar region • Continents flooded by shallow seas, limestones deposited	• Gondwana-land still in the southern hemisphere • Eurasia and North America lie in the tropics • Shales, slates and Old Red Sandstone laid down as continents flood during warm, dry climate	• Gondwana-land joins the northern land masses to form Pangaea • Warm climate cooled to an ice age in the southern hemisphere • Marine limestones (Mississippian) overlaid with coal seams (Pennsylvanian)	• Pangaea heads northward • Climates generally cooler • Shales, limestone, clay and sandstone and Red Beds of desert sandstone laid down • Much mountain-building	• Sea levels very low • Graywacke, shale, siliceous and red desert sandstone deposits • Arid climate over most of the land • Pangaea begins to split • Atlantic Ocean opens up	• Gondwa land contin break up • Graywac shale and siliceous sediments f • Seas floo much of the • Continue mountain-building • Mild, mo climates ov most of the
TERRESTRIAL ANIMAL LIFE	• No life on land?	• No life on land?	• Scorpions, millipedes and possibly eurypterids come out of the water	• Flightless insects, spiders and the first amphibians	• Age of Amphibians • First reptiles appear • Snails, centipedes and millipedes, cockroaches and giant dragonflies	• Reptiles begin to take over from amphibians • Mammal-like reptiles, lizards and others • Land invertebrates, particularly insects such as beetles and bugs, flourish	• Reptiles dominant, including the ancestors of dinosaurs and mammals	• Reptiles dominant w crocodiles, turtles, liza and dinosa • Pterosau *Archaeopte* take to the • Mamma begin to div • More advanced i such as flies
PLANT LIFE		• Primitive algae and seaweeds	• Primitive psilopsid plants live at edges of water	• Primitive, large spore-bearing plants form the first forests • Horsetails, clubmosses appear	• Rich flora of giant tree-ferns, horsetails and conifers form the swampy coal forests	• Giant clubmosses, seed ferns and horsetails • Pines and firs appear	• Gymnosperms such as ferns, cycads, gingkos and conifers	• Cycads, gingkos, co ferns and tr ferns
SEA LIFE	• Primitive algae and seaweeds • Jellyfish, sponges, starfish, worms, velvet worms • Shelled animals appear • Trilobites and brachiopods dominant, reefs built by archaeocyathids • First chordate lancelet	• Shelled invertebrates develop • Corals, bryozoans, brachiopods, bivalves, gastropods, nautiloids, trilobites, echinoderms, graptolites • Jawless armored fishes	• Marine invertebrates, crinoids, brachiopods, trilobites, nautiloids and graptolites • Reefs built by tabulate corals • Eurypterids (sea scorpions) in brackish waters • Fish became abundant in seas and fresh water	• Horn corals, brachiopods, ammonoids, crinoids, conodonts and trilobites • The Age of Fishes, jawed and jawless bony fish, armored fish, lobe-fin fish and cartilaginous fish all abundant	• Ammonoids, brachiopods in the seas • Rugose corals, graptolites, trilobites and some forms of bryozoans, crinoids and mollusks disappear	• Probably one-third to one-half of all marine families of animals become extinct by the end of this period, the biggest mass extinction ever	• Ammonoids dominate the seas • Bivalves and brachiopods are locally abundant but generally rare • All other major invertebrate groups are rare or missing	• All major invertebrate groups wel represente ammonoids bivalves successful • Marine reptiles, ichthyosaur plesiosaurs

	CENOZOIC ERA - 65 mill. yrs.					QUARTERNARY PERIOD – 1.5 mill. yrs.		
ACEOUS)D 136 million ago	PALEOCENE EPOCH From 65 million years ago	EOCENE EPOCH From 54 million years ago	OLIGOCENE EPOCH From 38 million years ago	MIOCENE EPOCH From 24.5 million years ago	PLIOCENE EPOCH From 5 million years ago	PLEISTOCENE EPOCH From 1.6 million years ago	HOLOCENE EPOCH 10-25,000 years ago	
s flood half land at esses of the shells of -celled ils, laid d masses to move d their nt positions nate mild ithout nes	• Seas retreat • South America isolated	• North America and Europe separate • Mountains rise • Seas flood the land • Warm climate	• Australia separates from Antarctica • Climate cools	• Ice covers Antarctica • Sea levels fall • Himalayas, Rockies and Andes rise	• Continents in their present positions, climate cools	• Ice ages cover northern lands • Sea levels fall	• As today, but many animals and plants becoming extinct	GEOGRAPHY AND CLIMATE
vanced aurs such as -bills tles, snakes, nanders lls and ng birds ossums other mals dinosaurs many other reptiles ct by end of d	• Age of Mammals • Rodents, early primates, herbivorous and carnivorous mammals • Large flightless birds	• Whales and sea cows, rodents, bats, early horses and elephants, lemurs and tarsiers • Marsupial mammals in South America	• Grazing mammals become more diverse, dogs, rats, pigs, hyrax, giant sloths	• Mammals at their most diverse • Sabre-toothed cats, monkeys and apes • Migrations of elephants from Africa to Eurasia • Cats, cattle, giraffes and pigs migrate the other way • Marsupials and monotremes in Australia	• Grazing hoofed animals successful • Mammals mostly as they are today • Human ancestors appear in Africa • Rats arrive in Australia	• Wooly mammoths and rhinos, saber-toothed cats, cave lions • Giant marsupials in Australia • Human hunting skills develop, many large mammals disappear	• Human civilizations begin	TERRESTRIAL ANIMAL LIFE
mnosperms, pias and sses wering s appear, olias and	• Flowering plants become more successful		• Grasses and trees cover much of the land	• Extensive grasslands	• Grasslands replace many forests in Africa			PLANT LIFE
nkton, coral rudists, onoids, reous algae rine reptiles mmonoids ct by end of d	• Gastropods, bivalves, foraminifera, fish, sharks successful		• Toothed whales replace early whales					SEA LIFE

THE "TREE OF LIFE"

It is possible that all living things today had one common ancestor, although this may not have been the first living thing to appear on Earth. Nature has carried out many experiments along the way, some successful and others less so. This "evolutionary tree" shows the main groups of living things according to the modern five-kingdom classification. Fossils can be fitted onto the various branches. Sometimes, finding an unusual type of fossil means creating a new "branch" for the tree.

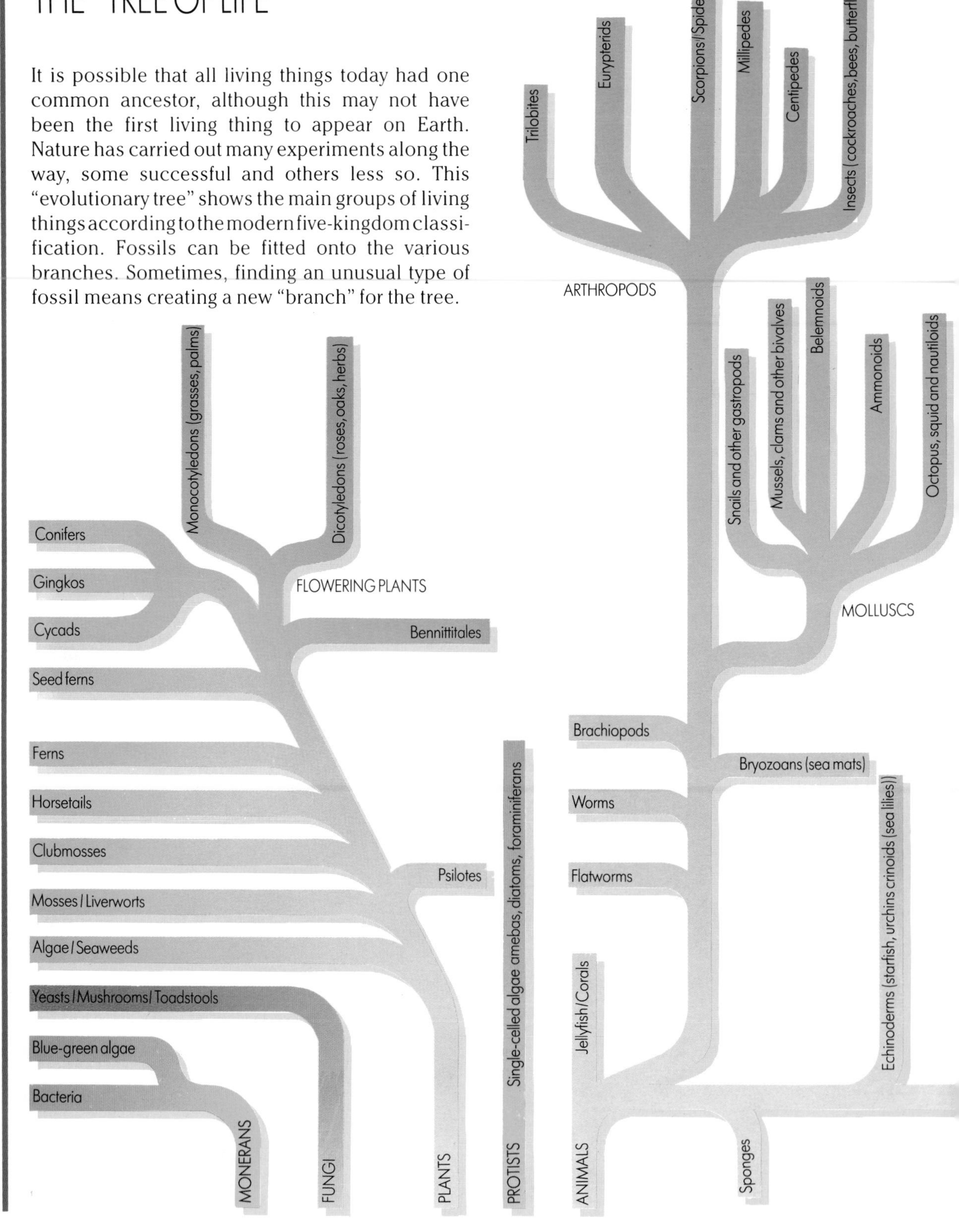

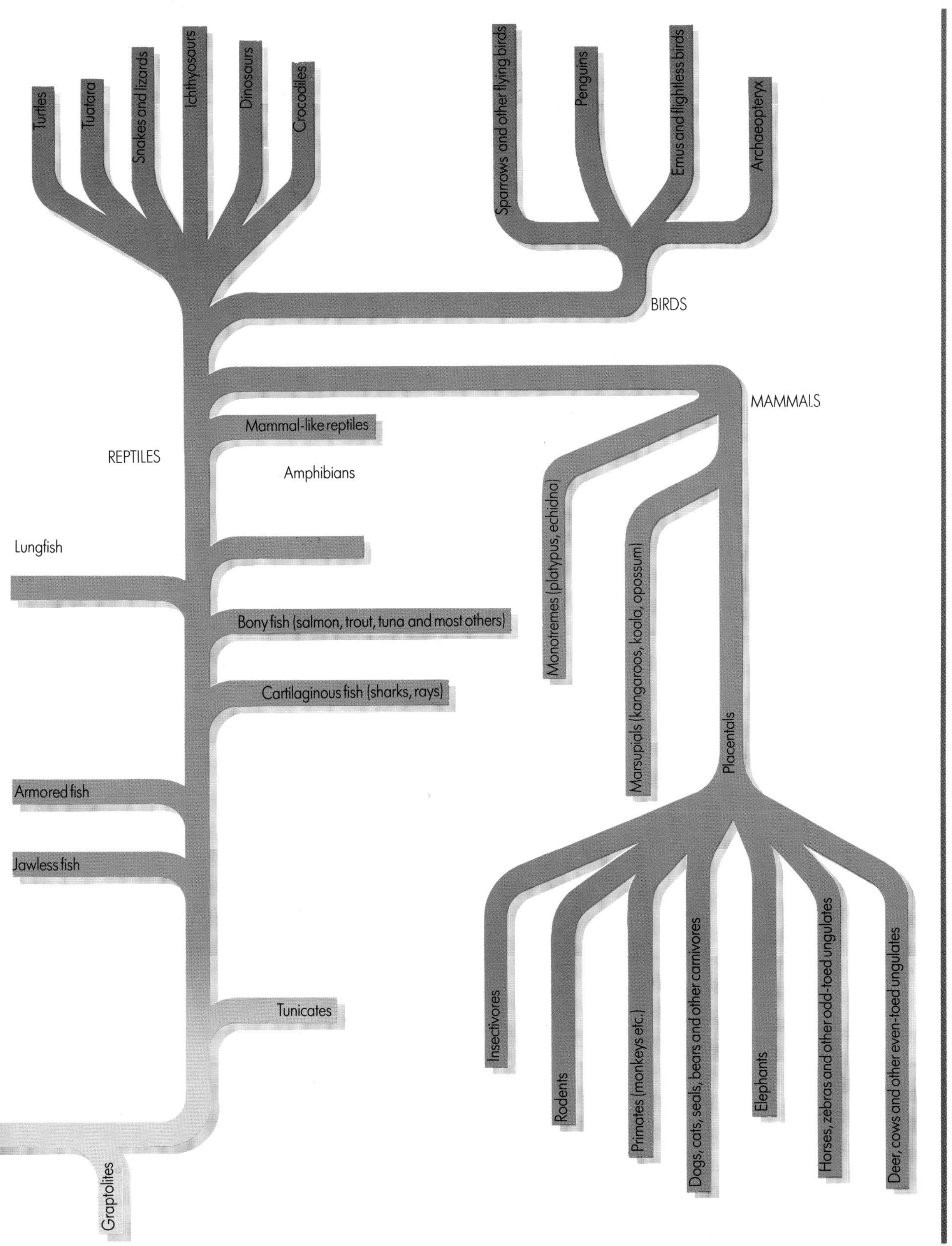
Turtles
Tuatara
Snakes and lizards
Ichthyosaurs
Dinosaurs
Crocodiles
Sparrows and other flying birds
Penguins
Emus and flightless birds
Archaeopteryx
BIRDS
MAMMALS
Mammal-like reptiles
REPTILES
Amphibians
Lungfish
Bony fish (salmon, trout, tuna and most others)
Cartilaginous fish (sharks, rays)
Monotremes (platypus, echidna)
Marsupials (kangaroos, koala, opossum)
Placentals
Armored fish
Jawless fish
Tunicates
Graptolites
Insectivores
Rodents
Primates (monkeys etc.)
Dogs, cats, seals, bears and other carnivores
Elephants
Horses, zebras and other odd-toed ungulates
Deer, cows and other even-toed ungulates

PRECAMBRIAN TO SILURIAN

Geography, geology and climate

The Precambrian era lasted from the beginning of the planet Earth, about 4,600 million years ago, until some 570 million years ago. It therefore accounts for about 87 percent of our planet's history. Most types of rocks are represented in the Precambrian strata, the oldest found so far being around 3.9 billion years old. In fact, the majority of all rocks date from the Precambrian, but they are usually so old and distorted that they are hard to interpret. There are sediments, indicating that erosion began soon after the first rocks formed. But the best clues, fossils, do not appear until 3,500 million years ago, and even then they are very rare and indistinct. It was not until the Cambrian period that the first "explosion" of life took place.

Clues in the rocks suggest that major events occurred billions of years ago as they have more recently: continents drifted, mountains were built, and glaciers gouged out huge channels.

The Precambrian era is divided into two parts, the Archean (or Archaic) period and the Proterozoic period. The Archean lasted for about 2.1

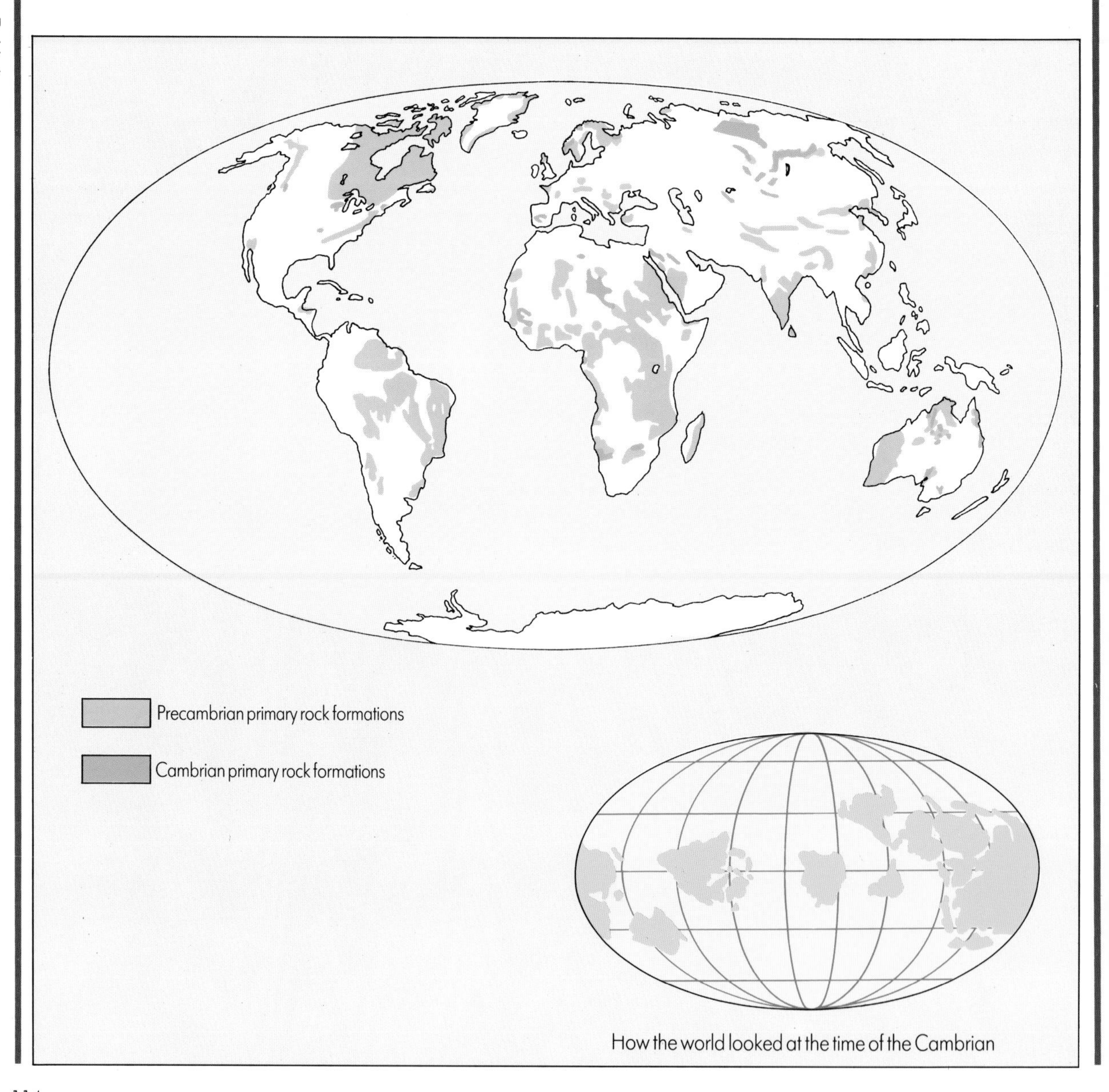

How the world looked at the time of the Cambrian

Precambrian sandstones from Greenland. The cliffs are of concretions in bedded sandstone, and the screes and slopes in the background are of the same age.

billion years, and during this time the Earth's crust and atmosphere developed. At first, the atmosphere consisted mainly of gases such as nitrogen and carbon dioxide – there was very little oxygen.

The seas were also in existence during this period. In fact, they have existed for as long as the continents, although their origins are unclear. Most of the elements that existed were dissolved in the sea, in varying quantities, having been washed from the land by rivers, or having flowed up from hydro-thermal springs on the seabed.

During the Proterozoic period, the simple bacteria and early algae thrived in the seas, and began to influence the physical world. Algal photosynthetic activity produced oxygen, which built up in the atmosphere. By the end of the period, the surface of the world was much as we know it today – but for the lack of life.

The beginning of the Cambrian period, about 570 million years ago, also marks the beginning of a new era, the Paleozoic. During the early Cambrian, conditions allowed the rapid evolution of life forms in the warm, shallow, mineral-rich waters. The main types of sedimentation were deep-water muds and shallow-water calcareous beds around the land masses.

During this period, the land itself was mainly desert, with sandstone deposits underlying some of the limestone. Most of the southern parts of what are now the continents of America, Africa, Asia and Australia collided to form the "supercontinent" of Gondwana, with much resultant mountain-building.

The Ordovician period, which started 500 million years ago, involved much volcanic activity and further mountain-building as the continents drifted into the southern hemisphere. What is now the Sahara Desert experienced glaciation as it traveled to the South Pole. Sea levels fell, then rose to their highest point. Extensive marine sediments of sandstone were overlaid with limestone.

The Silurian period began 430 million years ago with an ice age; its legacies are sediments of shales and limestones, especially dolomitic. The continents were again flooded, to such an extent that very little sand could be eroded from the land to form sandstone deposits. However, toward the end of the period, the seas receded.

Stromatolites

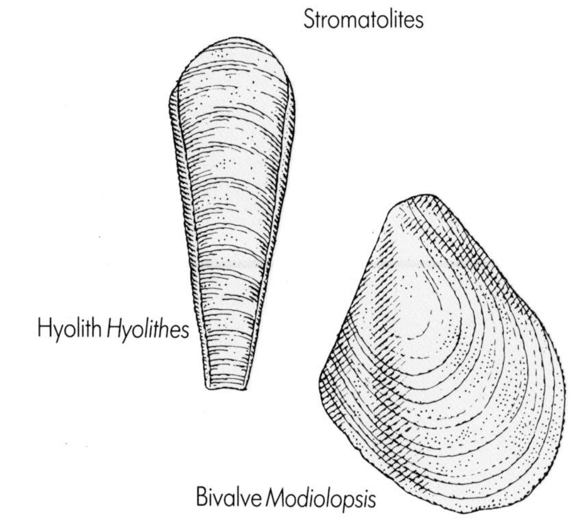

Hyolith *Hyolithes*

Bivalve *Modiolopsis*

This page A collection of invertebrate fossils ranging from Precambrian to Ordovician. The stromatoliths are the earliest structures in this group.

Early forms of life

The first living things probably appeared in the warm Archean seas. How, and when, and even why, are subjects for another book.

The earliest fossils are of single-celled, bacteria-like organisms, in rocks 3.5 billion years old. Blue-green algae, the first photosynthesizers, appeared some 300 million years later. Stromatoliths are structures formed when mats of algae are overlaid with sediment, grow through the layer, and are again covered. This process continues, to create a "layered lump." Stromatoliths occur in oceans today, but their fossils go back 2.9 million years and provide one of the few ways of dating early rocks.

Soft-bodied organisms do not fossilize well, so many clues to early life are missing. We cannot trace in detail how the first multicellular organisms evolved, how they lived and fed or how they bred. There are occasional glimpses of the life that did fill these primeval seas, such as the impressions of worms and jellyfish left in the sandstones which now form the Ediacara Hills of South Australia, dating from the Upper Precambrian, and the famed fossils of the Burgess Shale Beds in British Columbia, formed in the Middle Cambrian. Some of the groups represented in these glimpses of early life do not reappear in the fossil record; others, such as *Precambridium sigillum,* show the segmentation which later appeared in the arthropods.

There are many groups of soft-bodied worms in the world today, and they are especially important in the processes of decay and recyling of nutrients. Their presence throughout the fossil record is only hinted at by wriggling tracks, burrows, tunnels and casts. They probably existed in Precambrian times, and it seems likely that both arthropods and mollusks developed from types of soft-bodied worm.

Hard-bodied forms

Animals with shells and skeletons appeared at the beginning of the Cambrian. They are often already well developed at the time when their first known fossils appear in the record, so their evolutionary roots were probably earlier, in Precambrian times.

The lower multi-celled animals (*Metazoa*) developed hard parts. Sponges are supported by needle-like silica spicules, and their remains form rocks such as flint; they occur throughout the fossil record. The first reefs were built up by the skeletons of archaeocyatha, a group which became extinct before the end of the period. The coelenterate animals still have soft-bodied forms, such as jellyfish, but those that developed skeletons – the corals – have been the main reef-builders to the present day. Bryozoans, another group associated with coral-building, grow as branching colonies formed by the skeletons of countless individual organisms. They, too, form part of the fossil record from the Cambrian to the present.

Many hard substances have occurred in shells and skeletons: silica, phosphates, calcium carbonate and chitin (the protein-based covering of insects) are all examples. The reason why so many animals suddenly developed shells is unclear, but there may have been some changes in the chemical environment, such as in the acidity of the sea. The survival of animals with shells and skeletons is obvious today: it seems to have been the same then, and organisms with shells and skeletons

(Continued on page 120.)

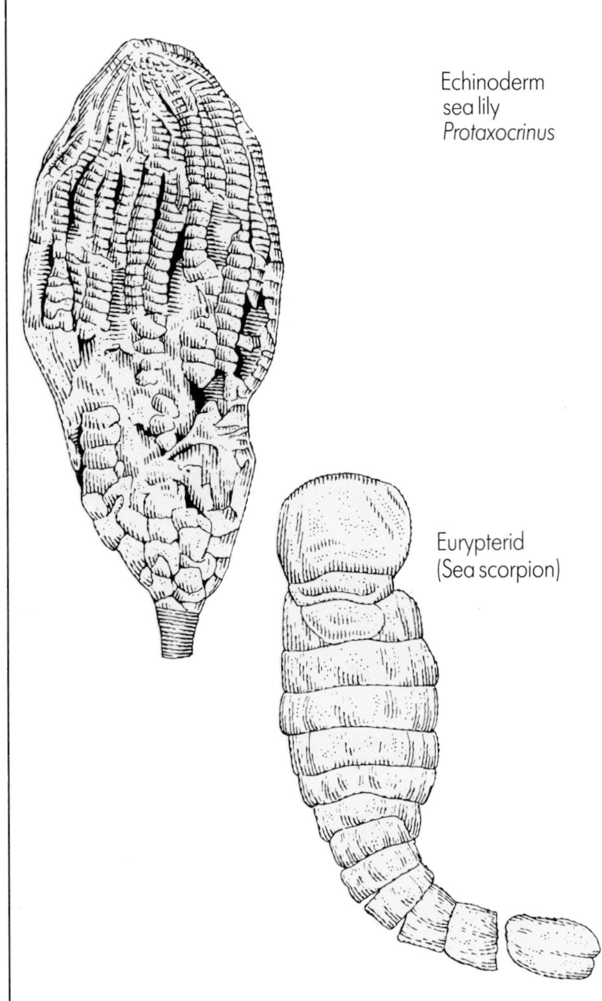

This page The early echinoderms appeared in the Cambrian period. Eurypterids spanned the periods from Ordovician to Permian.

Before the vertebrates

The life forms that existed before the beginning of the Cambrian era, 570 million years ago, left only very rare traces in the rock. These are usually found in very fine sediments that have been pushed up from their seabed origins, or as microfossils in very ancient rocks; they are unlikely finds for the amateur fossil hunter. But with the beginning of the Cambrian era, when hard-shelled animals entered the scene, fossils become more abundant. It would seem that their evolution was already under way before this time, since they appeared simultaneously in many and various forms. Through the next 200 million years, to the Silurian period, invertebrate animals representing all of today's major groups developed in numerous ways – many of which we may never know about. But the fossils they have left provide a tantalizing glimpse of life in those primeval seas.

(**The photographs shown here are approximately life-size.**)

Turret-like eyes

Above *Dictyonema flabelliforme*, a graptolite another early example of shelled animals. Each worm-like animal secreted and lived in a hard tube joined to its neighbor, forming branched colonies of many individuals. For 200 million years, from the middle Cambrian period, they thrived in the world's seas. Their branching patterns are characteristic of the different species.

Shell articulation

Right *Harknessella vespertilis*, a brachiopod from Ordovician rocks. Brachiopods superficially resemble mollusks such as oysters, but they are a separate group. Their two shells are arranged above and below the body, whereas mollusk bivalves have shells on each side of the body. The design has remained unchanged for more than 550 million years. Today's representatives, such as the lampshell *Lingula*, have a good claim to being the "oldest living fossil."

Single-row rhabdosome

Left *Monograptus triangularis*, a graptolite from Silurian rocks. Monograptolite colonies grew in single rows, without branching. They became more successful than the branched types and are the most common graptolites found in rocks of the Silurian and lower Devonian, the point at which they approached the end of their evolutionary history.

Left *Chaetetes corrugatus*, a sponge found in Silurian limestone. Sponges have lived on the ocean floor from the beginning of the Paleozoic era to the present. Most have an internal lattice-like skeleton made of siliceous spicules and a bony substance called spongin. They have a simple, bag-like, double-layered body which evolution has fashioned into many different shapes.

Left *Ogygiocarella debuchia*, a trilobite from Ordovician rocks. Trilobites were the first successful arthropods (animals with a chitinous outer exoskeleton and jointed legs). They appeared with the first shelled animals at the beginning of the Paleozoic era, and by the Ordovician they were enjoying their heyday. They dominated the seas, from the muddy bottom to the clear waters above.

Below right *Poleumita rugosa*, a gastropod (snail-like) mollusk from Silurian limestone. Mollusks are another group of shelled animals that appeared at the end of the Precambrian era and have enjoyed continuing success. Gastropods have only one, usually spiral, shell; their bodies are bent so that the tail is near the head. With this unlikely posture, they have conquered almost every habitat on the planet.

Left *Halycites catenularis*, a coral from Silurian limestone. "Chain corals" bloomed in the Paleozoic era but then became extinct. They lived in warm, slow-moving, shallow water on the tops of reefs. Each "link" of the chain is the calcareous skeleton made by an individual animal, similar to an anemone.

began to evolve and multiply at a great rate. Single-celled animals and plants such as foraminiferans, radiolarians, nummulites and coccoliths made themselves skeletons, or *tests*, and were so numerous that entire strata of sedimentary rocks which formed on the seabed are composed of their skeletons. They are useful stratigraphic indicators because of their great abundance and variety.

Brachiopods are among the most common fossils of the Paleozoic and Mesozoic eras. A few still exist today, as the lampshells. In fact, some forms, such as *Lingula,* have changed very little since the Cambrian. They have evolved to cope with most marine environments. The two shells, always unequal, are made of a chitin-phosphate substance. The articulated brachiopods are the most common types and, as their name suggests, their shells have a complicated articulation (joint). In fossils of these creatures, the two halves are nearly always still linked together.

Mollusks developed in the Cambrian from monoplacophoran forms. They have limpet-like shells, still represented today by *Neopilina,* and they probably gave rise to the other groups of mollusks. There were gastropods (snail-like forms) in the seas from early Cambrian times, their spiral

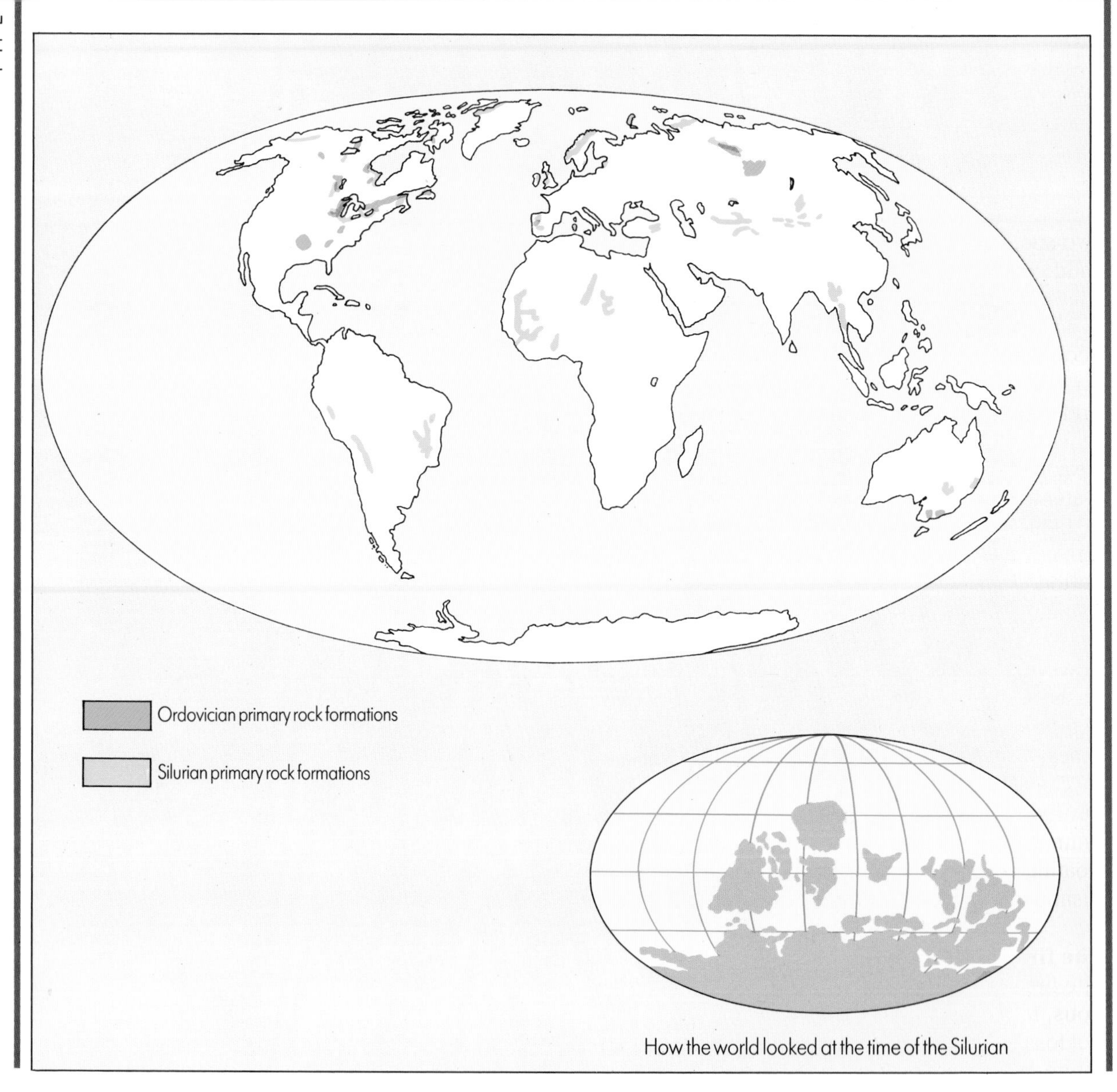

How the world looked at the time of the Silurian

shapes being many and varied. Bivalve mollusks also appear at the beginning of the Cambrian, but how they came to have two shells instead of one is unclear.

The cephalopod mollusks did not appear until the end of the Cambrian, when nautiloids came on the scene. They began as straight tapering shells. They were often very small although some were as large as 4 m (13 ft). The straight forms make up an important part of the Ordovician fossil record although they disappeared at the end of the Silurian, while the coiled forms, like the chambered nautilus, lived on to the present day.

Arthropods are represented throughout the Paleozoic era, including the crustaceans. They have an external jointed skeleton made of chitin, which is shed at various stages during growth; the majority of arthropod fossils are therefore of cast shells. Trilobites, for example, were so numerous and of such variety that they are the predominant index species for Cambrian strata. Ostracods, tiny two-shelled crustaceans, have been buried in the mud on the sea floor since Ordovician times, and they too are important index species.

Other arthropods, loosely related to spiders and scorpions, are less numerous, but more spectacular. The eurypterids appeared in the Ordovician and reached 10 ft. (3 m) and more in length. These fearsome carnivores prowled the brackish waters near the shore, and some may even have left the water for short periods.

Echinoderms – starfish and their relatives – were already abundant by the end of the Cambrian period. They have calcareous plates in their skin usually exhibiting a radial symmetry based on five or its multiples. Some of the earliest groups, however, show no symmetry, or a radial symmetry based on two or three. Crinoids (sea lilies) and echinoids (sea urchins) appeared in Ordovician times, although they become more important later.

The main index fossils of the Ordovician and Silurian periods are the graptolites. These colonial animals lived attached to the seabed or to some floating mass; others had floats of their own. They disappeared in the Carboniferous period.

The first vertebrates

The first creatures with "backbones" had cartilaginous, rather than bony, skeletons, and so they did not fossilize well. The first traces of the bony plates found in primitive fish appear in the fossil record some 500 million years ago. More complete remains have been found in the Lower Ordovician, of the small 5 in. (12 cm) *Arandaspis*. It had a bony carapace at the head end, rows of scales along its sides, but no fins or jaws. The first jawed fishes evolved in Upper Ordovician; they reached their zenith in the Devonian "Age of Fishes."

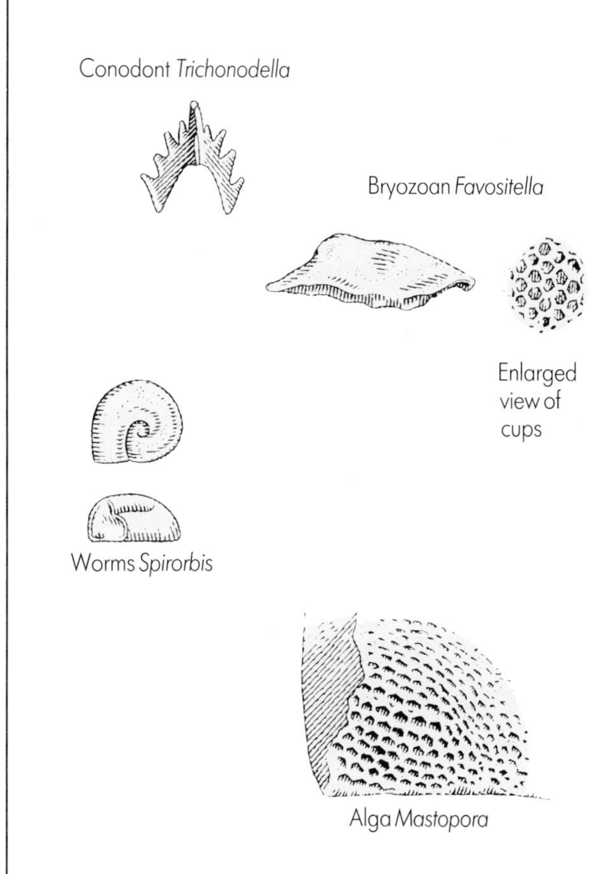

This page Paleozoic fossils. Bryozoans (sea mats) are colonies of tiny, individual, coral-like animals that form calcareous cup-shaped shells. They first appeared in Ordovician seas, and they are still found today on rocks and seaweed fronds, often looking like "filigree" lacework.

THE UPPER PALEOZOIC ERA – DEVONIAN TO PERMIAN

Geography, geology and climate

During the later (or "Upper") Paleozoic era, from the beginning of the Devonian period (3.95 billion years ago), through the Carboniferous period (3.45-2.8 billion years ago), to the end of the Permian (2.25 billion years ago), the continents continued drifting around the globe. North America and Europe slowly collided to form the continent of Laurasia in the Devonian period, and the Caledonian mountains appeared at this time across the new continent. Their remains are now found in North America, Britain and Scandinavia. This mountain-building involved much metamorphism, weathering, erosion, and deposition of sediments. The famous Old Red Devonian sandstone produced this way is found in Scotland, New York, Pennsylvania and West Virginia, in layers up to 13,000 ft. (4,000 m) thick. The oceans again flooded the continents, then receded.

During the Devonian period, oxygen in the air reached levels which would have provided ade-

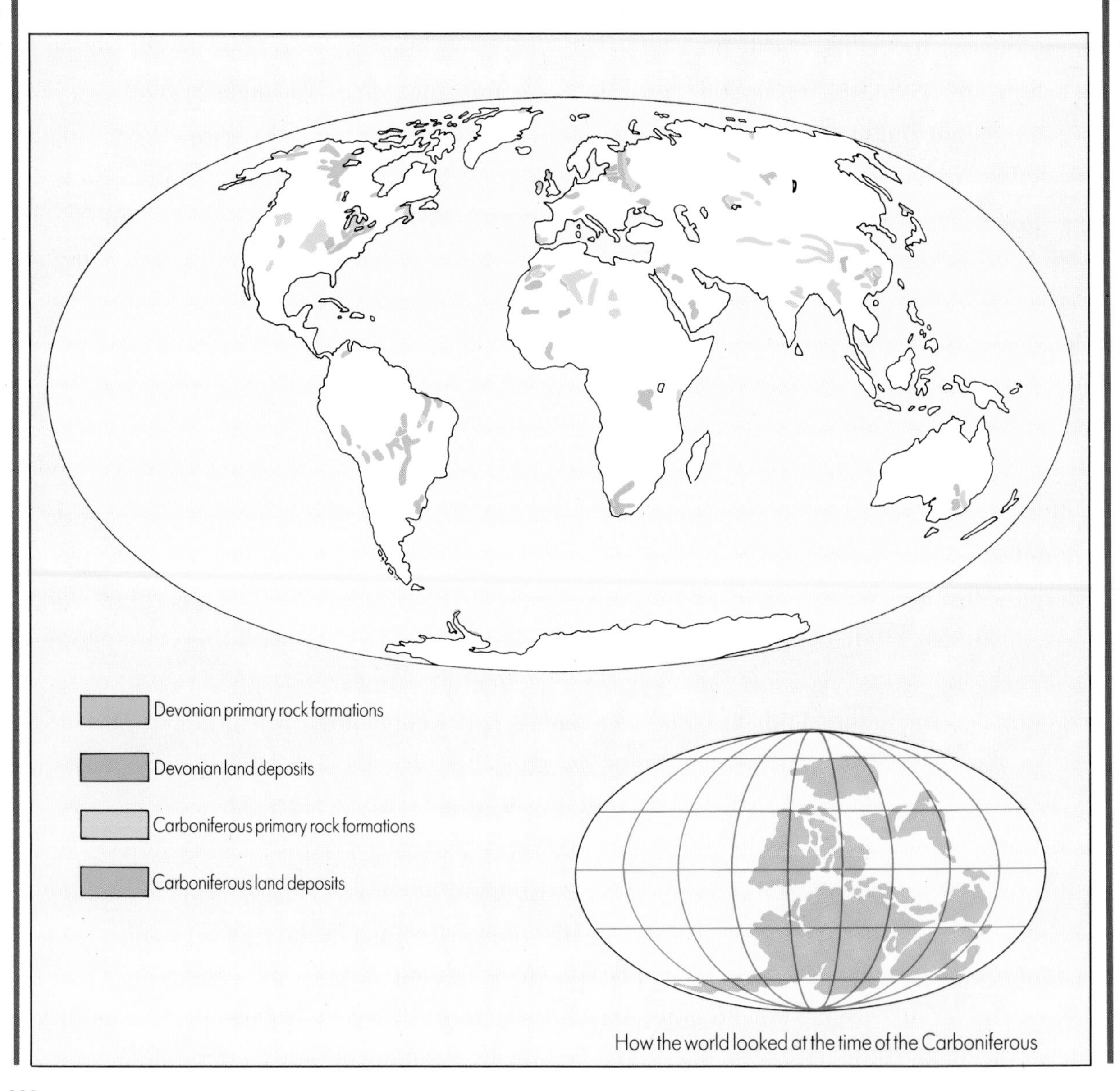

How the world looked at the time of the Carboniferous

These sandstones and mudstones were laid down beneath rivers and lakes toward the end of the Permian period, just before the end of the Paleozoic Era. The Karroo plateau in South Africa is constantly worn away by torrential rains and baking heat, except where the soft rock is protected by volcanic basalt. The area is rich in fossils of freshwater creatures.

quate protection, as an ozone layer, from the Sun's ultraviolet radiation. Life could not have escaped from the cover of the water before this time.

In the Carboniferous period, Gondwanaland moved northward toward Laurasia. There was an initial ice age in southern parts, while the northerly regions were warm and moist – providing ideal conditions for the swampy tropical forests that quickly sprang up. Sea levels rose, and gray and blue-gray limestones and shales were laid on the Devonian Red sandstones. In the U.S.A., such lower Carboniferous sediments are known as Mississippian. Tropical and subtropical forests on the swampy deltas were flooded as the sea rose, then grew again, forming the great coal deposits that characterize the later part of the period, known as Pennsylvanian.

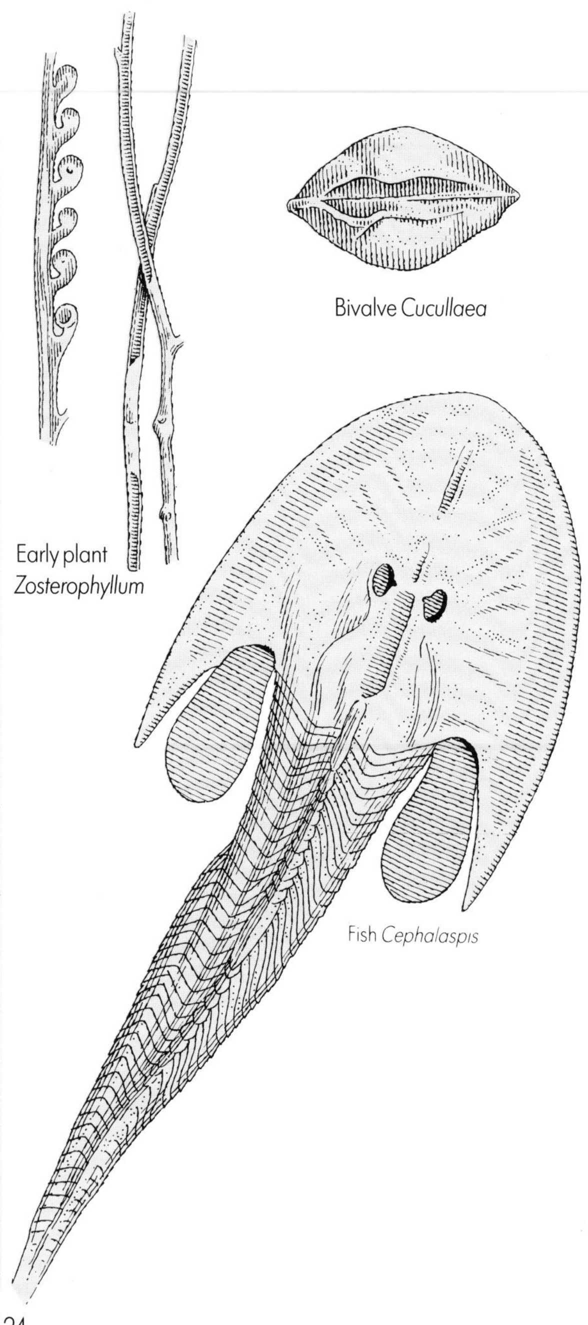

Bivalve *Cucullaea*

Early plant *Zosterophyllum*

Fish *Cephalaspis*

The collision between Laurasia and Gondwanaland formed the supercontinent of Pangaea during the Permian period. There was much uplifting of land, and ranges such as the Appalachians, Rockies, Alps and Urals rose at this time. The new single continent allowed the rapid dispersion of land animals and plants. There was again much erosion from the new mountains, laying down sandstones in the sea, which today form the Red Beds of Texas, North America and Africa.

The greatest-ever mass extinctions in the sea and on the land, with representatives from every group being lost, mark the end of the Paleozoic era. This event coincided with the formation of Pangaea, but its causes are not clear. Large quantities of evaporites were deposited in rocks, and the extinctions may have been associated with changes in the sea's salt concentration and/or temperature.

Plants

The Devonian is known as the "Age of Ferns and Fishes." Plants had existed in the water from earlier times, and seaweeds must have been common, although their remains are not.

In the water, plants have nutrients and respiratory gases dissolved all around them, and they need no internal means of support, being buoyed up. Land plants faced new problems. They needed stiff, supporting tissues to keep them upright; roots to absorb moisture and nutrients from the soil; and a vascular system to transport water and nutrients around the plant's tissues. They also need protection from drying out, and new means of spreading their spores and seeds.

The first vascular plants had appeared in the Silurian period, but there was little development until the Devonian. The first land plants were leafless and rootless, and grew around the edges of swamps and springs. Later came plants with roots, such as scale trees, tree-ferns and horsetails. They grew to great size and produced the first forests.

During the Carboniferous period, the run-off of erosion products from the land formed deltas and swamps. They were ideal for the development of lush tropical and subtropical forests. Giant tree-ferns, seed-bearing ferns and horsetails grew abundantly. Their remains form coal seams.

In the Permian, land plants were decimated by glaciation. The scale trees and horsetails declined, having been replaced by conifers (firs and pines), as well as the seed plant *Glossopteris* and the cycads.

Invertebrates

In the Devonian period, the first limestone sponges appeared, building great reefs. Graptolites gradually went into decline, but other invertebrate groups continued to diversify, leading to an evolutionary surge during the Carboniferous. Shales of this time contain crinoids, corals and mollusks. Early ammonoids developed and replaced many nautiloids. Yet after this surge, in the Permian, representatives of many groups were lost. Trilobites disappeared completely. Sea levels fell and drained the continental shelves of the shallcw, warm seas in which invertebrates thrived. Ammonoids, however, continued to evolve successfully.

Several groups of arthropods attempted to conquer dry land. Their exoskeletons provided support and protected them against drying out, and the burgeoning land plants offered an untapped food source. Millipedes may have been among the first of these land-dwelling arthropods, feasting on rotting plant remains; some grew to nearly 6 ft. (2 m) in length. Arachnids came on the scene at this time, with ticks sucking the sap of the earliest plants, and spiders appearing to hunt them. Scorpions nearly 3 ft. (1 m) long also lived in the leaf litter of these early forests. The first insects, bristletails and springtails, appeared at this time and remain almost unchanged to the present.

The great Carboniferous coal forests were ideal for insects, which soon conquered the air and grew to enormous sizes compared to present-day species. For example, there were giant dragonflies with 2 ft. (65 cm) wingspans. In the Permian, various beetles, bugs and cicadas appeared. Another invertebrate group that mastered life on land, the gastropod mollusks, made their debut as land snails crawled through the leaf litter.

Vertebrates

As the land was being populated by plants and arthropods, the vertebrates followed them in search of food. Devonian fishes developed from bottom-living, jawless varieties (agnathans) to jawed types, with either cartilaginous or bony skeletons. Freshwater habitats were colonized. The early Osteostraca, from the Red Devonian sandstone, were small and had a solid bony carapace; the bone was thick and molded around the internal organs. Such fossils are well preserved, and much is known about the animal's internal

Continued on page 128.

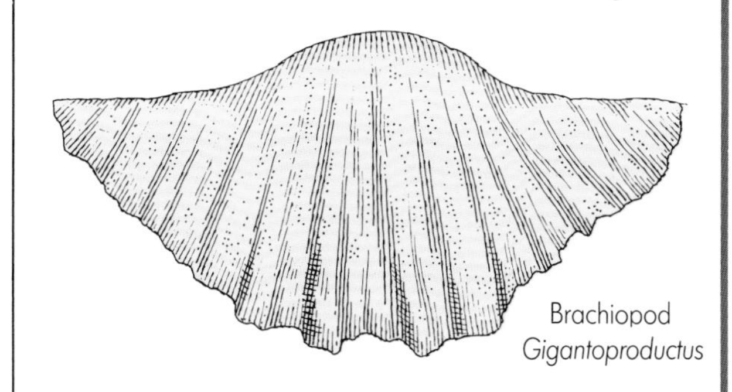

Brachiopod *Gigantoproductus*

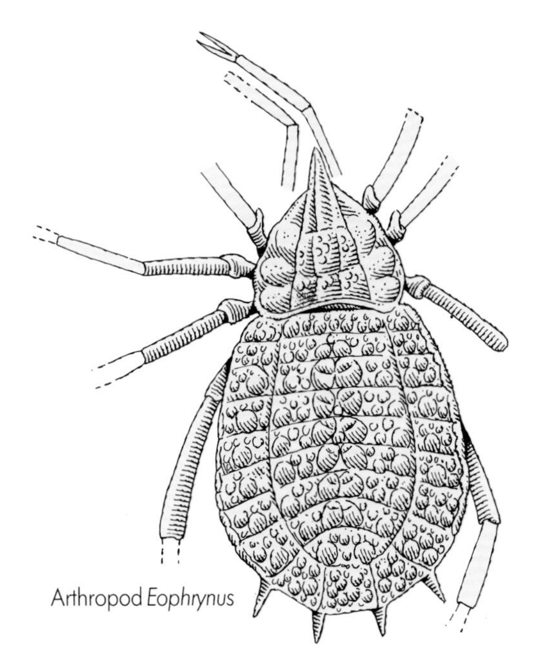

Arthropod *Eophrynus*

Opposite page and above Middle and Late Paleozoic fossils. Vertebrates such as the first fish have appeared in the sea, and the land is being colonized by plants and small animals, particularly millipedes, scorpions and other arachnids, and insects. *Eophrynus* was the second fossil arachnid to be discovered, and was described by William Buckland in 1836.

Conquest of the land

During the latter half of the Paleozoic era – the Devonian, Carboniferous and Permian periods – evolving plants, invertebrates and then vertebrates made their own bids to conquer the land. Fossils of organisms that survived on dry land are not so common, since water-borne sediment is virtually a prerequisite for fossilization. The swamps, lakes and rivers of the time are the main sources of fossil material. Life continued to evolve in the seas, however, and there is an abundance of material from uplifted marine sediments of this age. By the end of the era, the forerunners of many major groups of plants and animals were represented.

(**The photographs shown here are approximately life-size.**)

Fern fronds

Above *Corvaspis kingi*, an ancient fish from the well-known Old Red Sandstone rock formations of the Devonian period. This particular formation, the Upper Downtonian, outcrops at Corvedale in Shropshire, England – hence the scientific name of the fossil.

Above *Mariopteris nervosa*, a pteridosperm or seed-fern from a Carboniferous seam. Seed ferns were large, woody plants with fern-like leaves. They were among the first plants to bear true seeds, which resembled the seeds of today's conifers. Because the seeds had no protective layers, these plants are called gymnosperms or "naked seeds."

Below *Osteolepis major*, a lobe-fin fish from the Old Red Devonian Sandstone. Fishes with fleshy fins appeared 390 million years ago. Coelacanths and lungfishes still survive but the group to which *Osteolepis* belonged disappeared before the end of the Paleozoic era; however, their legacy was the tetrapod group that evolved from the same stock.

Right *Annularia stellata*, a horsetail from Carboniferous coal seams. Horsetails grew almost everywhere on land during the period, and this species was one of the most advanced. These plants grew very tall, but their trunks contained pith, not true wood. They reproduced by means of spores released from cones.

Below *Tealliocaris woodwardi*, a shrimp from Carboniferous deposits. Fossilizaton of decapod ("ten-footed") crustaceans was fairly rare and required very fine, quick-settling sediments. This group of arthropods originated in Permian times, but did not flourish until relatively recently.

Left *Reticuloceras reticulatum*, a goniatite mollusk found in Carboniferous shales. Many types of ammonoid mollusks (the group which includes the ammonites) came and went during the Paleozoic era, and the goniatites were among the early ones.

Right *Amphoracrinus atlas*, a crinoid or sea lily from Carboniferous times, when these starfish-related animals were at the height of their success. There were many different types and their body plates, "horns" and horizontal arms fossilized well.

Right *Lonsdaleia floriformis* coral from Carboniferous rocks. Corals have been forming reefs since the end of the Precambrian era. Each species is very particular about where it lives on the reef, and it is possible to study today's species and then estimate the depth of the water, its temperature and its nutrient richness at the time.

structure and how it relates to modern forms. Anaspids, less bony and resembling modern lampreys, are also found in the same deposits.

Jawless vertebrates had a restricted diet, since they were only able to suck and grind their food. With the appearance of jaws, there came a new form of predation. Acanthodians, early jawed fish from the Ordovician, flourished in freshwater habitats. They had paired fins, and were capable swimmers, with the large eyes of a hunter. Soon the huge, armored placoderms were reaching lengths of 22 ft. (7 m), with fearsome blades in their jaws. The first sharks appeared, too, and their original design proved so successful for predation that it has hardly changed today. Sharks have cartilaginous skeletons, and thus their fossils are rare, except for their teeth.

Bony fish first appeared at the end of the Silurian period. They had complex skeletons and swim-bladders to maintain buoyancy. There were two main groups. One evolved to become the teleosts, which returned to the sea in Permian times and have dominated both salt and fresh water ever since. The vast majority of today's 20,000-odd fish

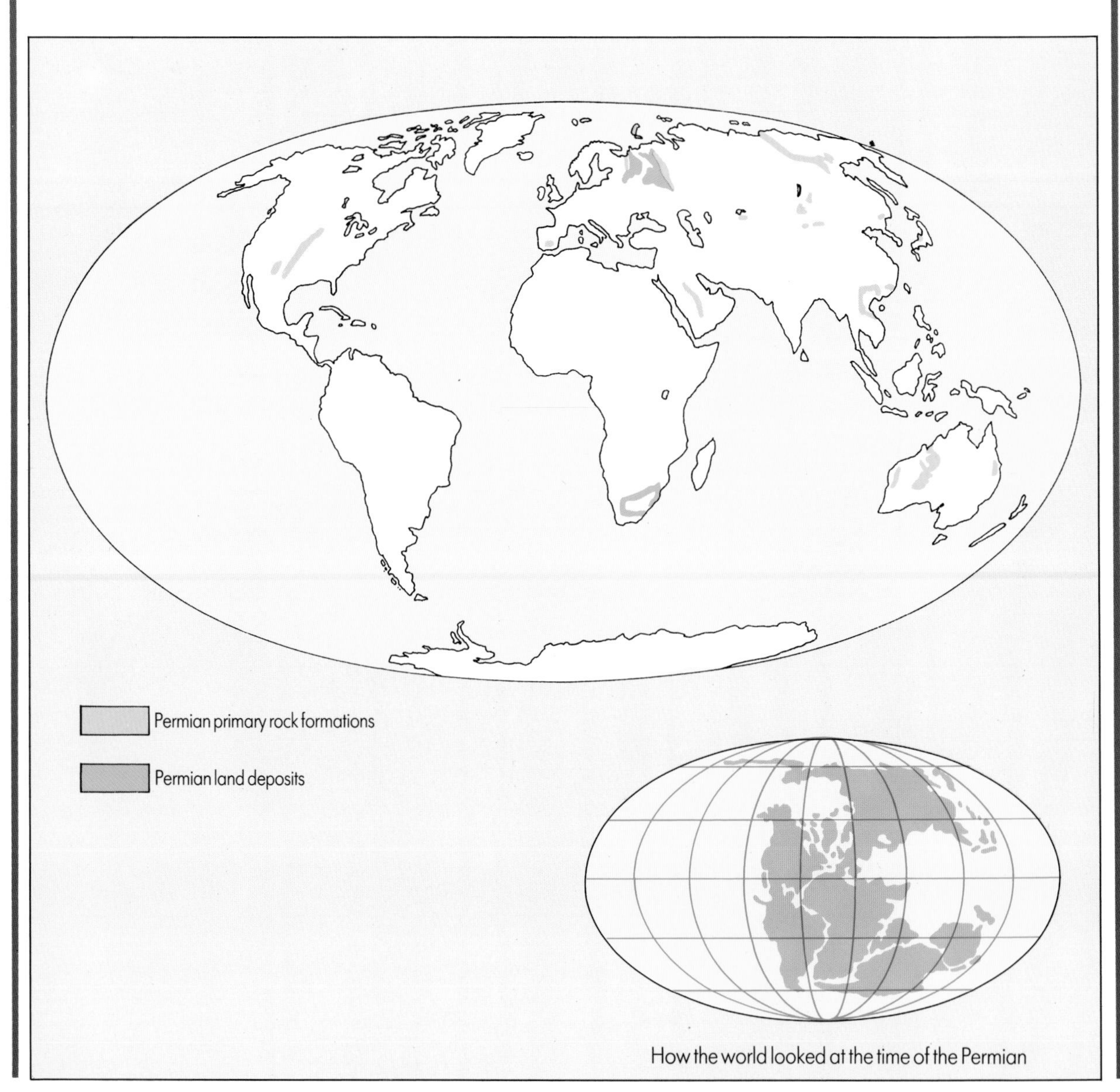

How the world looked at the time of the Permian

species, from minnows to pike, are teleosts. The other group were similar to the recently discovered coelacanth. They had lobe-shaped fins, and their ancestral group led both to lungfishes and to land vertebrates.

The Carboniferous period was the "Age of Amphibians." The bony structures inside the fishy lobe-fin had become the tetrapod leg, with five toes, and shoulder and hip girdles for support and articulation. Amphibians probably crawled from the water into the damp forests in search of invertebrate prey. They still laid eggs in water and had a tadpole stage in their life cycle, and to this day, they have not overcome their aquatic link. Even so, they evolved to fill every ecological niche in the Carboniferous forests. Some were sturdy and as big as a crocodile; others had tiny limbs, while still others soon lost their newly acquired limbs and slithered off snake-like through the leaf litter.

A second group of tetrapods began to tread the ground toward the end of the Carboniferous period and during the early Permian. The first reptiles were mainly small insect-eaters. They began to diversify considerably during the Permian. The two early groups were cotylosaurs, directly descended from amphibians, and pelycosaurs, distant forerunners of mammals.

During the Permian period, these reptiles became more widespread at the expense of the amphibians. Reptiles had at last broken the tie with the water. They developed the shelled, amniotic egg, which could be laid on dry land. They also acquired a scaly, leathery skin that resisted desiccation. They were ready to dominate the world.

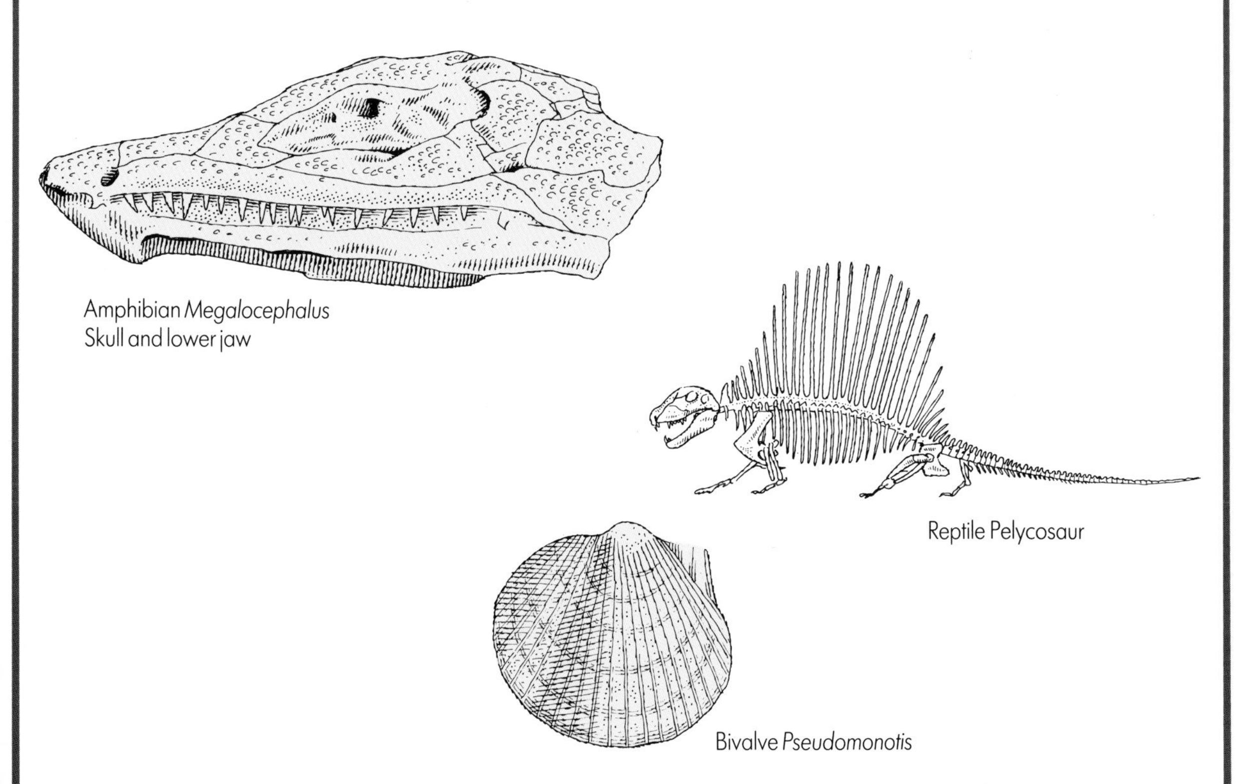

Amphibian *Megalocephalus*
Skull and lower jaw

Reptile Pelycosaur

Bivalve *Pseudomonotis*

This page Late Paleozoic fossils. The strangely crocodilian skull of *Megalocephalus* ("Huge-head") belies its amphibian status. Pelycosaurs were early reptiles with characteristic sails, probably used for temperature regulation. The ever-present bivalves continued to evolve in the seas.

THE MESOZOIC ERA – TRIASSIC THROUGH CRETACEOUS

Geography, geology and climate

During the Triassic period, which began 225 million years ago, the supercontinent of Pangaea remained as one. The interior of this great land was quiet, although there was some volcanic activity around the periphery; the climate was generally poor. Sea levels rose, then fell again by the end of the period, and consequently Triassic sediments are often in three bands: two red, land-derived sequences separated by a band of marine limestone, shale and sandstone. The red beds consist of sandstone, siltstone and shale, their color due to iron oxide which is indicative of an arid environment. There are also some minor coal seams.

In the Jurassic period (beginning 193 million years ago), Pangaea began to fragment. Laurasia split from Gondwana, resulting in the birth of the Atlantic Ocean. Gondwana itself split further, into three parts: Africa and South America; Australia and Antarctica; and India. These movements and ridge formations of the sea floor displaced water

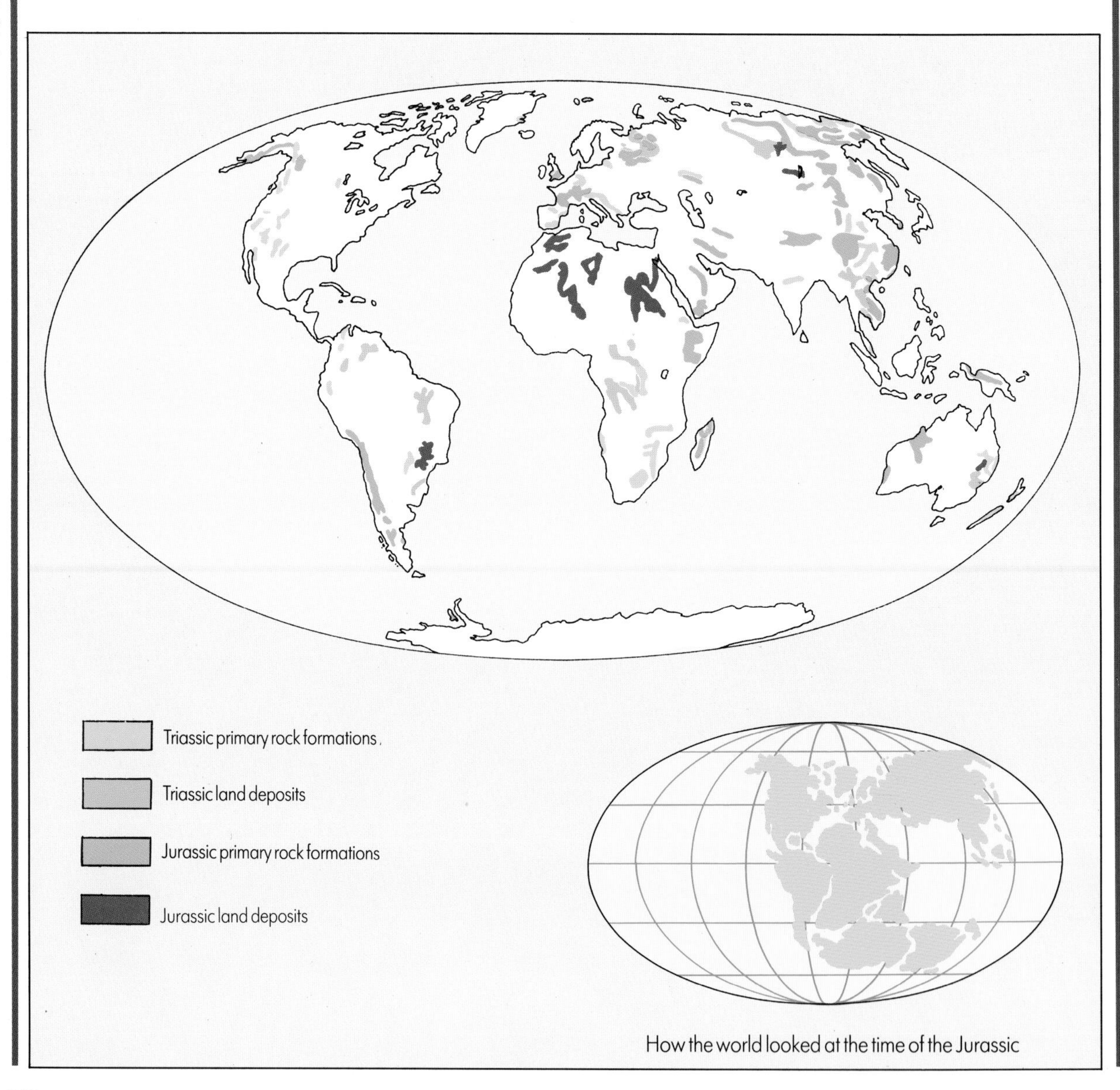

How the world looked at the time of the Jurassic

Many Mesozoic fossils have been found in the "badlands" country, seen here in a gulleyed landscape in the Dakotas. Visitors can almost trip over bones and teeth being eroded at the surface.

that caused sea levels to rise. A quarter of the continental area was flooded with warm, shallow seas. There was extensive sedimentation of marine shales, clays, limestones and sandstones, rich in fossils. The main deposits were limestones (chalk) formed from the ooze on the ocean floor, built up by the new calcareous-shelled planktonic organisms of the time. This ooze has been forming ever since the Jurassic, but it has been restricted to deep water since the end of the Mesozoic.

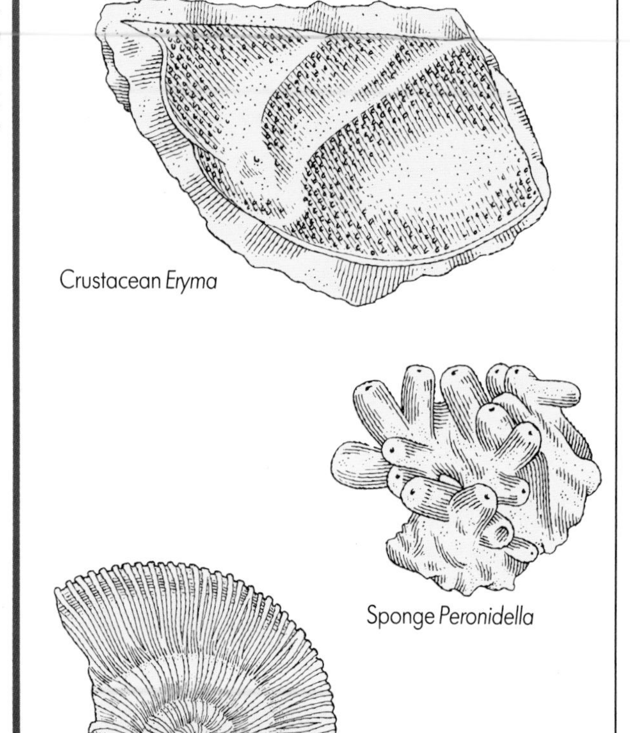

Crustacean *Eryma*

Sponge *Peronidella*

Ammonite *Kosmoceras*

Opposite page and above Mesozoic fossils. All the main groups of invertebrates were still evolving in the seas, the ammonites being very successful during this time. Bony fish are many and varied, as they are today.

This was the Age of the Reptiles, here represented by the crocodile *Steneosaurus*. Crocodiles, which belonged to the same group as the dinosaurs – the Archosaurs – have changed little to this day.

The Cretaceous period, which ended 65 million years ago, saw the continents inching their way toward their present familiar positions. At the start of the period, low sea levels meant sediments were derived mainly from the land. As the continents drifted, the ocean floor was again lifted by ridges, and sea levels began to rise again, reaching their highest levels thus far, with one-half of the land flooded. This formed the huge chalk sediments characteristic of this period. Some Cretaceous chalks are over 1,000 ft. (300 m) thick and often contain flint seams derived from sponges, as well as oil, gas, coal and metal ores. There was much mountain-building worldwide, with the early growth of many present-day chains such as the Alps, Andes and Himalayas.

The climate of Cretaceous times was mild, and the fossil record is abundant. Life flourished at all levels, but at the end of the Cretaceous came the mysterious mass extinction that killed off the dinosaurs and many other forms of life. Drastic changes in temperature, salinity, sea level, cosmic radiation, magnetic fields, oxygen concentration and even the stars have all been proposed as the cause. One clue is that sediments from the Cretaceous/Tertiary boundary often contain iridium, an element found in meteorites. The impact of a large meteorite would have thrown up dust clouds and darkened the skies for some time, interfering with plant photosynthesis, and having repercussions along the food chains.

Plants

The poor climate of the early Triassic meant a sparse flora, but the plant life was richer by the end of the period. The plants were mainly ferns, and *Glossopteris* was still abundant. Gymnosperms (plants with seeds that are not enclosed within a fruit) dominated the Jurassic. Cycad-like plants which bore seeds, ferns, horsetails, giant club-mosses and conifers, some resembling the modern "monkey-puzzle tree" (Araucaria), made up the forests. The gingko or "maidenhair tree" also appeared and has remained almost unchanged to the present.

An extremely significant step in the development of the plant world was the evolution of angiosperms, or flowering plants, toward the end of the Jurassic. They soon took over the land during the Cretaceous. Forests of deciduous, broadleaved trees emerged and angiosperms constituted nine-tenths of all fossil plant species by the middle of this period. Magnolia is one of the oldest flowering plants in the fossil record.

Invertebrates

Insect evolution closely followed that of the plants. Beetles were probably the first to take advantage of the nectar and pollen produced by flowers, since bees are not known until the Tertiary period. Most of the insects of Cretaceous times remain little changed today.

The extinctions at the end of the Permian mainly affected marine invertebrates. Lower Triassic marine deposits show a lack of fossils, with bivalves and brachiopods in abundance in only a few sediments (which, in fact, consist of little else).

However, later in the Triassic, life expanded to fill the empty seas. Marine invertebrates began to resemble modern forms, with mollusks in particular becoming dominant. One family of ammonoids survived the Permian extinctions and underwent rapid evolutionary change, becoming common as fossils and displaying various decorations, such as transverse ridges or increasingly complex tracings of the suture lines that divided their shells. Ammonoids are the main stratigraphic indicators of deep marine Mesozoic deposits. Bivalve mollusks also became important, as did new types of coral. Corals are very sensitive to, and therefore good indicators of, environmental conditions.

The majority of Jurassic sediments in the sea are of calcareous coccolithophores and planktonic foraminiferans. The latter, along with ostracods, are the main stratigraphic indices of shallow water. The Jurassic seas abounded with brachiopods, echinoderms, corals, belemnoids, and crustaceans such as shrimp and lobsters.

Cretaceous deposits are also mainly planktonic chalk. The seas contained huge reefs built by corals, bryozoans, stromatophorids, and calcareous algae. Sedentary clams called rudists, which resembled corals, lived among them. Ammonoids were still numerous, but suffered extinction along with many other forms at the end of the Cretaceous.

Vertebrates

The vertebrates were not so affected by the Permian extinctions, and diversified further during the Mesozoic era. Bony fish and sharks continued to develop gradually toward modern forms. Amphibians became less dominant by the beginning of the Triassic; most groups perished, leaving only the frogs and salamanders that we know today.

The Mesozoic was truly the "Age of the Reptiles." During this era, they came to dominate land, sea, and sky. There were running dinosaurs, swimming ichthyosaurs, flying pterosaurs, and the

(Continued on page 136.)

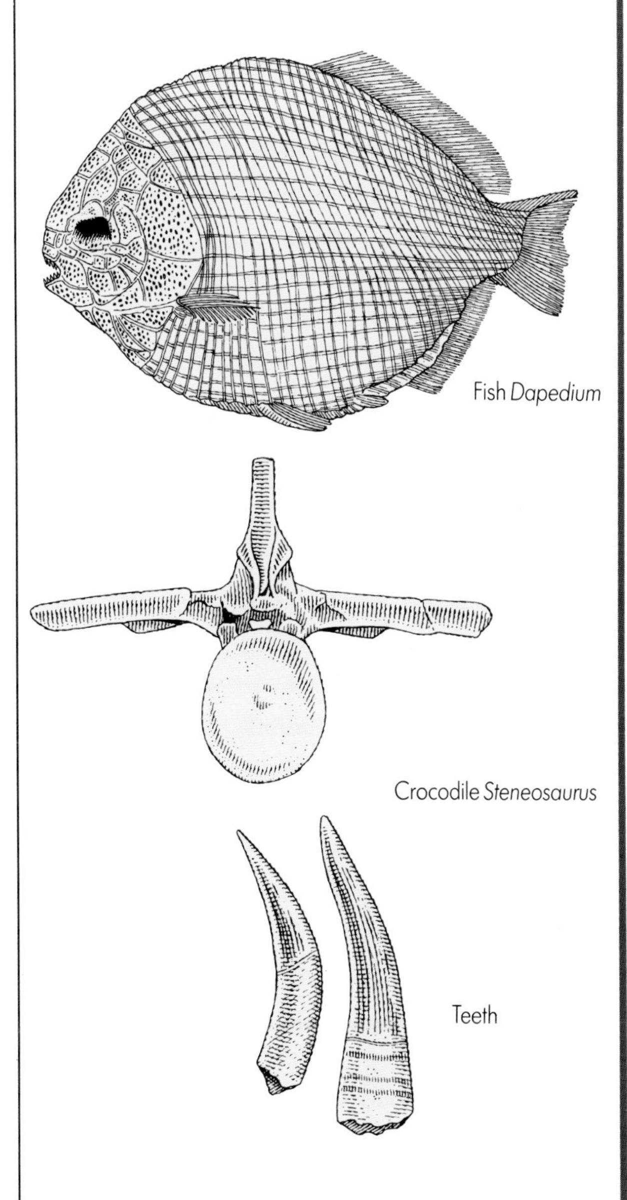

Fish *Dapedium*

Crocodile *Steneosaurus*

Teeth

Ruling reptiles

The fossil record of the Mesozoic era is dominated by the reptiles – and especially the dinosaurs, which have been capturing the popular imagination for over a century. But many other groups of animals and plants continued to evolve in the shadow of the dinosaurs. There were no dinosaurs in the sea or in the air; the swimming ichthyosaurs, plesiosaurs and mosasaurs, and the flying pteranodons, represented the reptiles in these habitats. Traces of the mammals, birds and plants that lived alongside the great dinosaurs are rare, due to their small size and consequent fragility. In the sea, major groups such as the mollusks continued their evolution.

(**The photographs shown here are approximately life-size**.)

Fan-shaped leaf

Above Plants, especially non-flowering ones, evolved rapidly during the Mesozoic period. Some have survived with little change until the present, although many others were overtaken by the flowering plants which appeared toward the end of the era. *Gingko digitata* was a gingko (maidenhair) tree from Jurassic oolite rocks. Gingkos first appeared 150 million years ago and became common in the Mesozoic period. They are gymnosperms and produce seeds without a protective covering. One species, *Gingko biloba*, survives as a "living fossil" today.

Central stem

Left *Coniopteris hymenophylloides*, a fern from Jurassic rocks. Ferns appeared in the Devonian period and are common fossils of Mesozoic sediments. They reproduce by means of spores and still grow in profusion today.

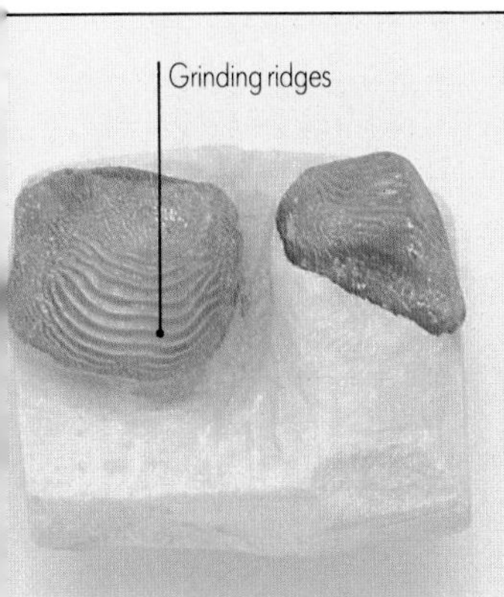

Left *Ptychodus decurrens*, a fish whose teeth were found in Cretaceous chalk. Fishes originated way back in the Devonian period, but it was in the Cretaceous era that bony fishes came to dominate all the world's waters, displacing others such as sharks and lobefins.

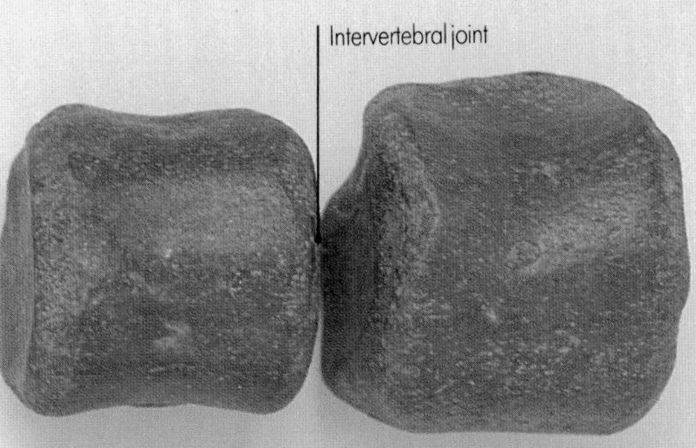

Below *Lopha colubrina*, an oyster from Cretaceous Greensand. Oysters are particular about their habitat and prefer warm, clean, moving water. The shell edges of *Lopha* are thrown into radial folds, the number being characteristic of the species. These folds serve to increase the area of the opening for the intake of water, without increasing the width of the slit.

Pointed end was at rear of animal

Left *Hibolites hastata*, belemnite shells from Jurassic clay. These fossils are also found in Asia, North America and Indonesia. These cephalopods are known to have had hooked suckers on their tentacles because their imprints have survived in the Solnhofen lithographic limestone that preserved *Archaeopteryx*.

Above The vertebra of a *pliosaur* from Cretaceous Greensand. These reptiles returned to the sea from a life on land and became the dominant carnivores of Mesozoic seas. Plesiosaurs (pliosaurs were short-necked versions of the flippered giants) are common fossils of Jurassic and Cretaceous rocks in Europe, North America and Australia.

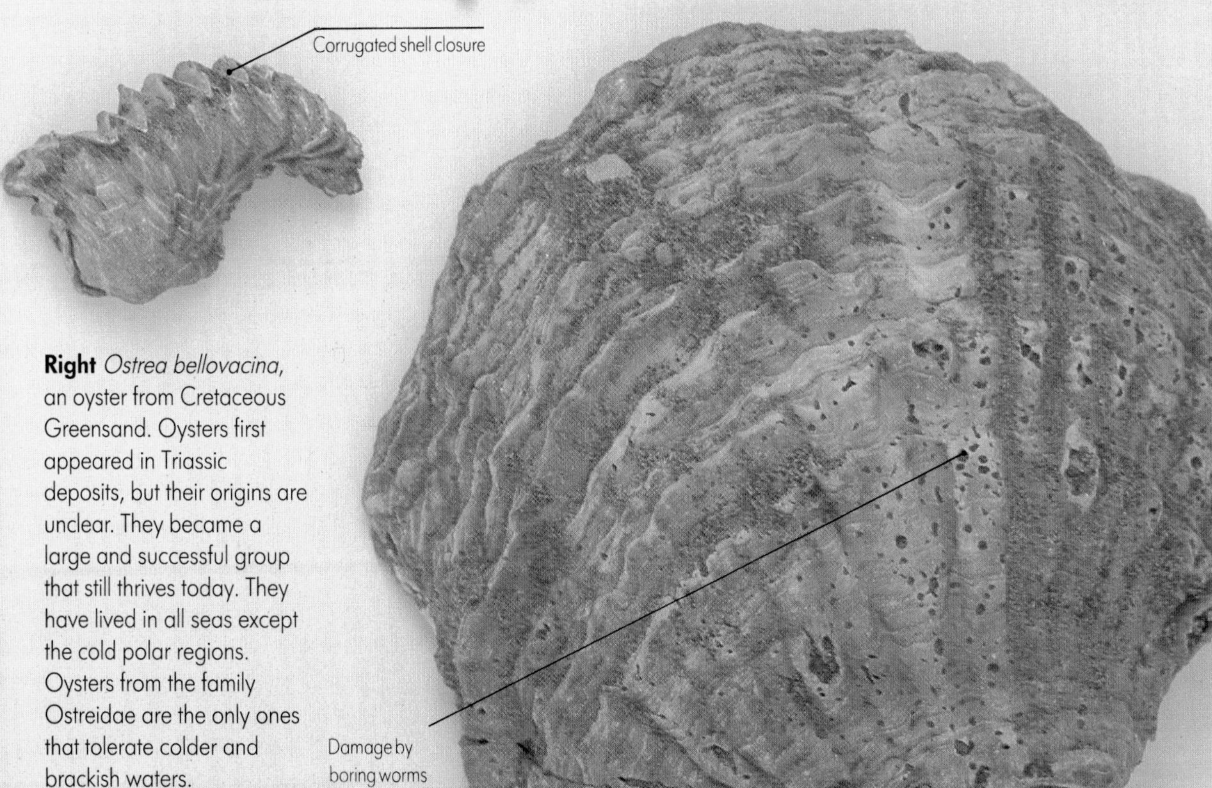

Right *Ostrea bellovacina*, an oyster from Cretaceous Greensand. Oysters first appeared in Triassic deposits, but their origins are unclear. They became a large and successful group that still thrives today. They have lived in all seas except the cold polar regions. Oysters from the family Ostreidae are the only ones that tolerate colder and brackish waters.

therapsids, which are linked to the evolutionary line that led to the mammals.

During the Triassic, reptiles such as *Placodus*, *Nothosaurus* and ichthyosaurs lived in the sea. On land were reptiles such as early lizards (some of which could spread skin on extensions to their ribs and glide), armored reptiles and a group now represented by the tuatara, a lizard-like reptile. During the Jurassic, pterosaurs dominated the air, while ichthyosaurs, plesiosaurs and turtles dominated the waters.

In the Cretaceous period, giant versions of plesiosaurs, mosasaurs, turtles and pterosaurs evolved. Crocodiles assumed their almost modern form. Snakes also developed, but their fragile fossils are rare.

It is unclear exactly when the therapsid-like reptiles became mammals. There are a few mammal fossils from the Upper Triassic. These tiny, shrew-like, insectivorous mammals continued to develop modestly, hiding from the great reptiles, through the Jurassic and Cretaceous. They had fur

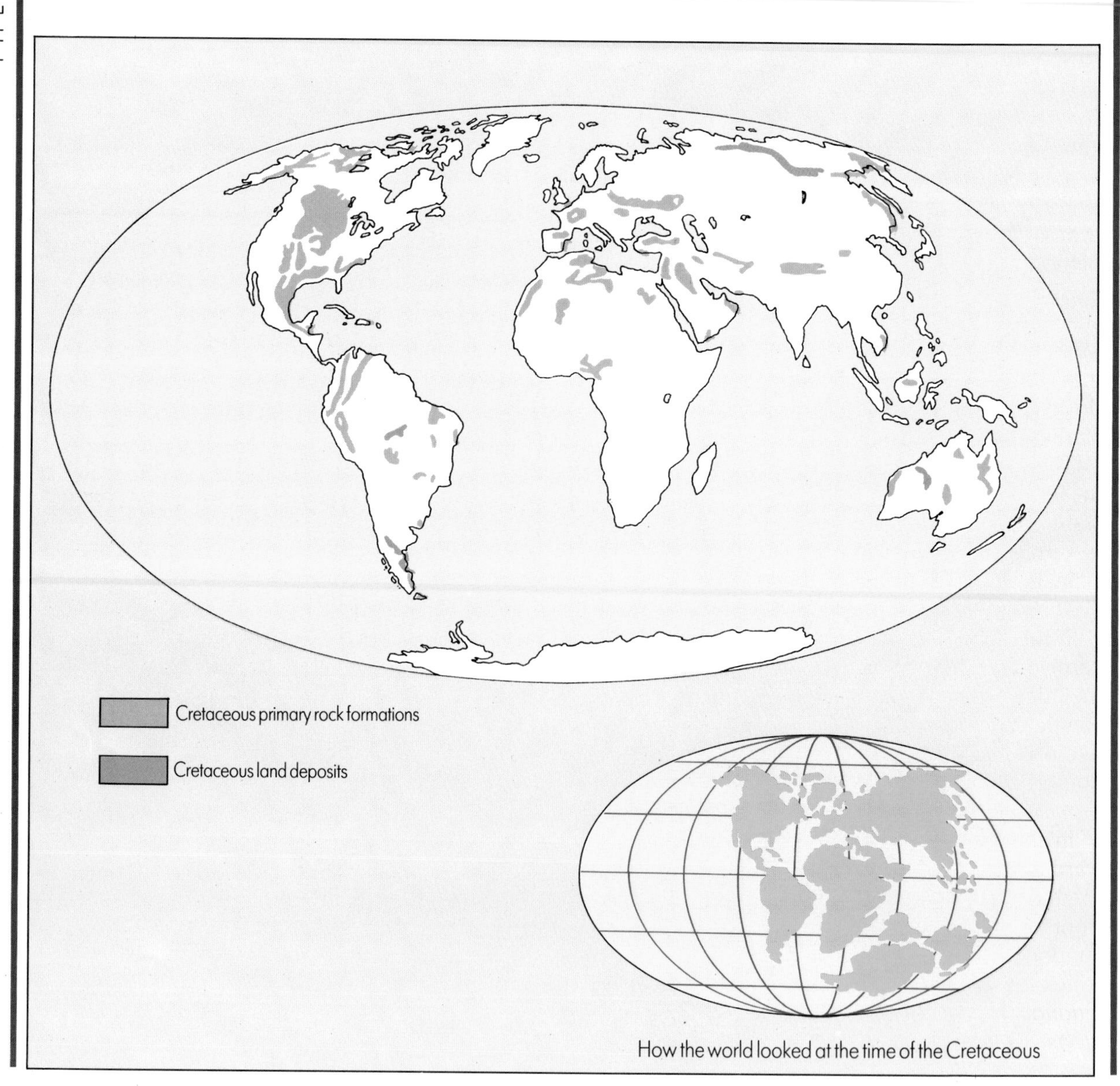

How the world looked at the time of the Cretaceous

(like some of the reptiles) and were probably warm-blooded. They could come out and forage at night, when the mainly cold-blooded dinosaurs were slow and inactive.

Another significant newcomer of the Mesozoic was *Archaeopteryx*. This bird-like reptile, or reptile-like bird, evolved from the same evolutionary stock as the dinosaurs. Its rare and beautiful fossils are dated as Jurassic, but it is not until the Cretaceous period that more birds appear. Their rare fossils include animals such as *Ichthyornis* and *Hesperornis*.

The dinosaurs

Dinosaurs came on the scene during the Triassic period. Their main early evolution occurred in the Jurassic; some forms disappeared in the Lower Cretaceous, but the group then underwent a resurgence, and there were many new species at the end of the era. None survived the mass extinction.

We know a great deal about dinosaurs – their evolution, anatomy, physiology, reproduction and behavior. Their bones were generally large and fossilized well. The fossils are common, and some skeletons are almost complete. These great creatures have also caught the imagination, and the thirst for knowledge about their size and habits drives paleontologists to further discoveries and studies.

The dinosaurs are divided into two main groups: those with hip bones arranged in a reptile-like fashion, the Saurischia, and those with a bird-like hip arrangement, the Ornithischia.

In the Saurischia, the pubic bone of the hip girdle hangs downward and forward. Dinosaurs with this feature in turn fall into two groups. The carnivorous theropods ran on two hind legs and included the huge and fearsome tyrannosaurs, the slender coelurosaurs, and the toothless, ostrich-like ornithomimosaurs. The herbivorous sauropodomorphs include giants such as *Diplodocus* and *Brachiosaurus*.

In the ornithischian hip, the pubis lies backward against the ischium. The Ornithischia were predominantly herbivores, with a bony pad on the front of the lower jaw, and a bony trellis-work supporting the spine. The pachycephalosaurs or "thick-headed" reptiles belonged to this group. *Iguanodon*, the stegosaurs, ankylosaurs, *Triceratops* and the "duck-billed" hadrosaurs represent its other branches.

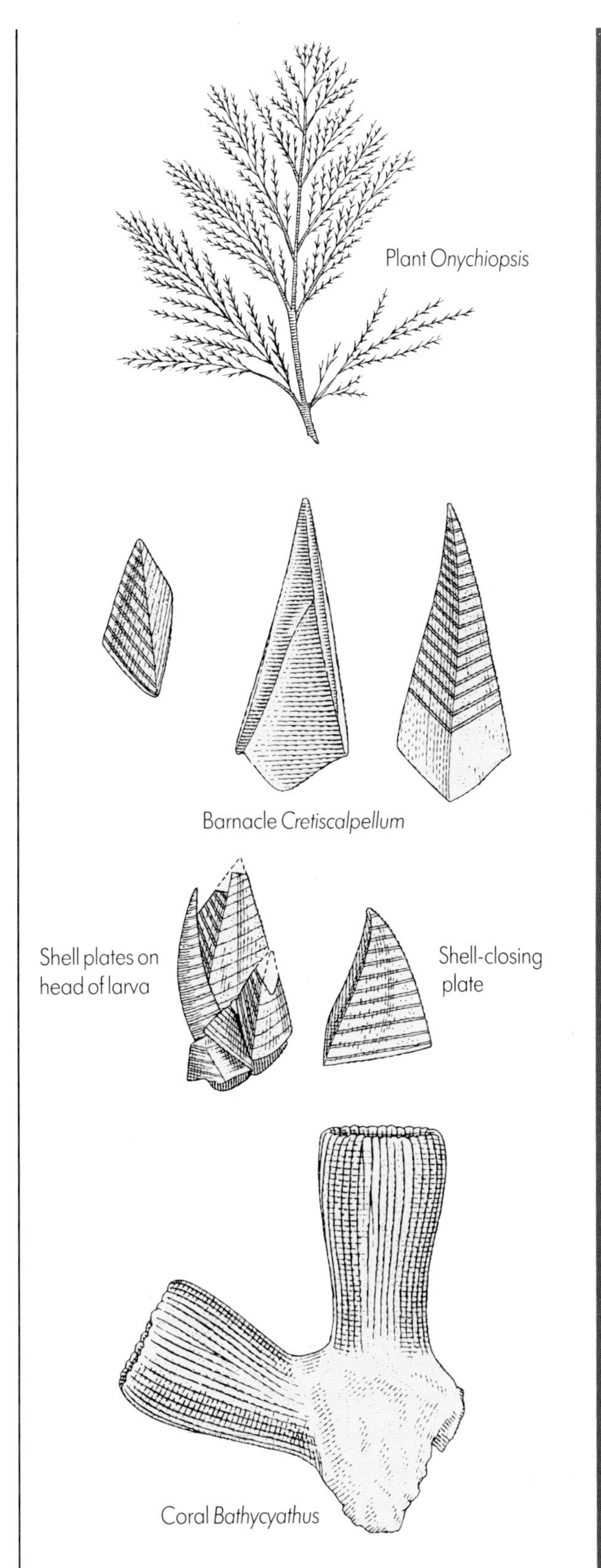

Plant *Onychiopsis*

Barnacle *Cretiscalpellum*

Coral *Bathycyathus*

THE CENOZOIC ERA – TERTIARY AND QUATERNARY

Geography, geology and climate

The Cenozoic era started 65 million years ago and is divided into two periods, the Tertiary and the Quaternary. During this time, the continents gradually assumed their present positions, and the global climate slowly became similar to that of today. At the beginning of the Tertiary period, it was warmer and moister, but the advent of the ice ages toward the end of the period caused temperatures to drop.

Tertiary rocks are formed mainly from marine sediments. They are often loosely consolidated and highly fossiliferous.

During the first epoch of the Tertiary – the Paleocene (65-54 million years ago) – India, Asia and Europe remained separate, while Australia was still joined to Antarctica. Sea levels fell again, and chalk deposits gave way to mainly continental-derived sediments. The South Atlantic was not fully formed; Europe and North America were still joined in the north, and consequently the Paleocene

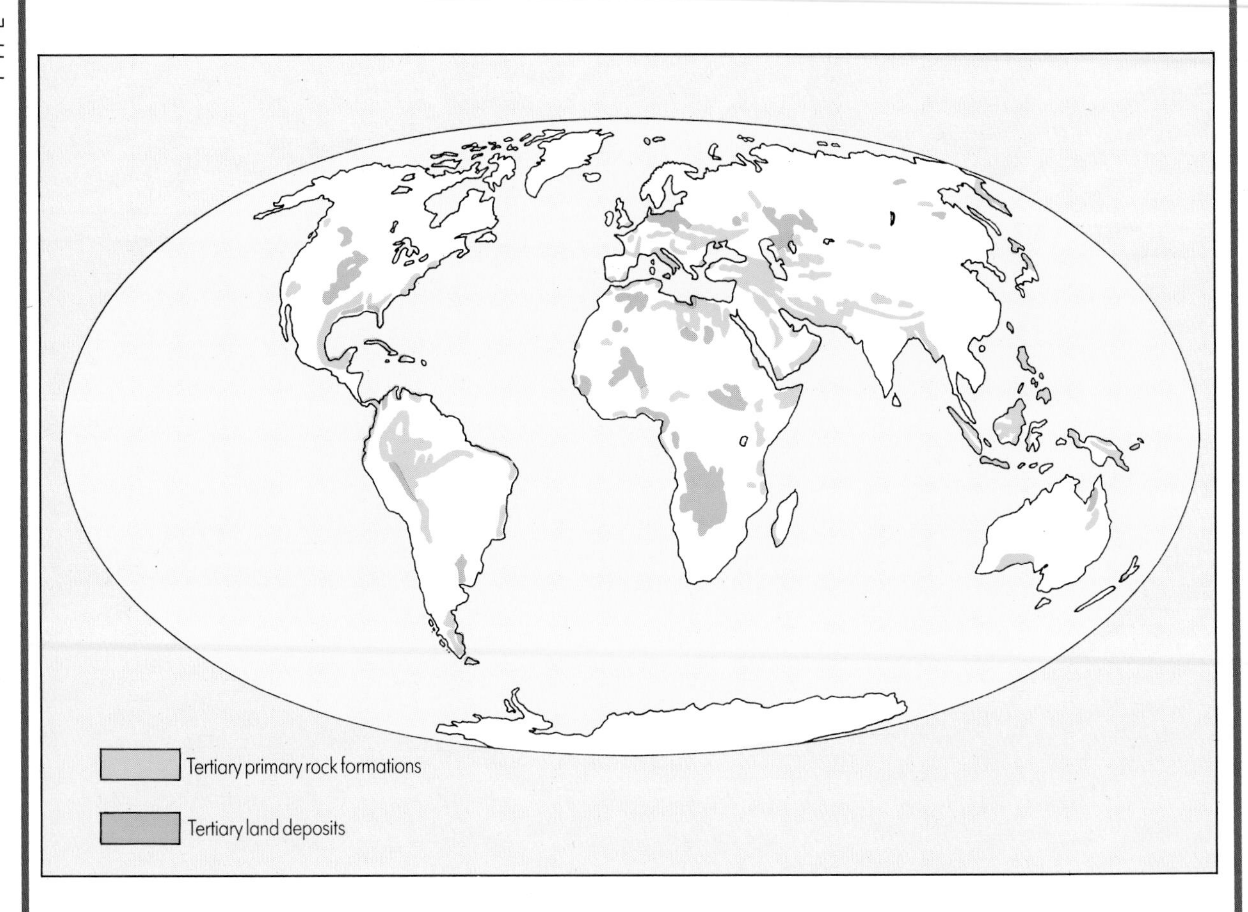

The River Cuckmere meanders through chalk deposits of the Eocene epoch in the South Downs of southern England. Fossils from the beginning of the "Age of Mammals," about 50-40 million years ago, have been found in this type of country.

animals of these two regions remained very similar.

During the next epoch, the Eocene (54-38 million years ago), the North Atlantic split Europe and North America, except when sea levels were exceptionally low. By this time, the first glaciation of Antarctica occurred, but the temperature gradients of the oceans were not as great as they are today. The world was still largely tropical with a warm, mild climate.

During the Oligocene epoch (38-26 million years ago), Australia finally separated from Antarctica. The currents that predominate in the oceans and affect global variations in climate were now established. Glaciation caused the climate to cool, and sea levels declined to their lowest. Sedimentation from this time is therefore limited. India joined Asia, and the collision made the Himalayas even higher. Europe joined with Asia, allowing the great animal migrations.

The Miocene epoch (25-5 million years ago) saw Africa approach Europe, leaving the Mediterranean Sea open, but pushing up the Alps. There were extensive ice sheets in Antarctica, and sea levels remained low.

During the last epoch of the Tertiary, the Pliocene (which ended some 1.6 million years ago), North and South America became joined at the Isthmus of Panama, so cutting the link for marine creatures between the Atlantic and Pacific. An ice sheet appeared in the Arctic, which marked the start of the recent ice ages.

Through the Quaternary period, which runs from 1.6 million years ago to the present day, the polar ice sheet spread south in waves, covering and then retreating from parts of Europe, North America and northern Asia, until the end of the Pleistocene epoch (about 10,000 years ago). The weight of the ice on the land caused the Earth's crust to sink, and the ice bound up so much planetary water that sea levels fell as the sheets advanced.

During the final epoch, the Holocene, the great ice sheets retreated to the poles. The climate became warmer, but more varied, from cold poles to the warm equator. The day-to-day weather became more changeable and stormy – the weather we have today.

Plants

Fossil plants from the Tertiary period are very similar to modern forms, having undergone most of their development during the previous Cretaceous period. The warm, wet swamps of the Eocene were filled with bald cypress and trunkless palms, while drier areas supported forests of magnolia and sabal palm. Gingkos, cycads, conifers such as cupressus and pine, and flowering plants like oak, plane-tree, maple, myrtle, alder and beech, all make their mark on the fossil record.

The decreasing temperature and more arid conditions caused the development of a new group of

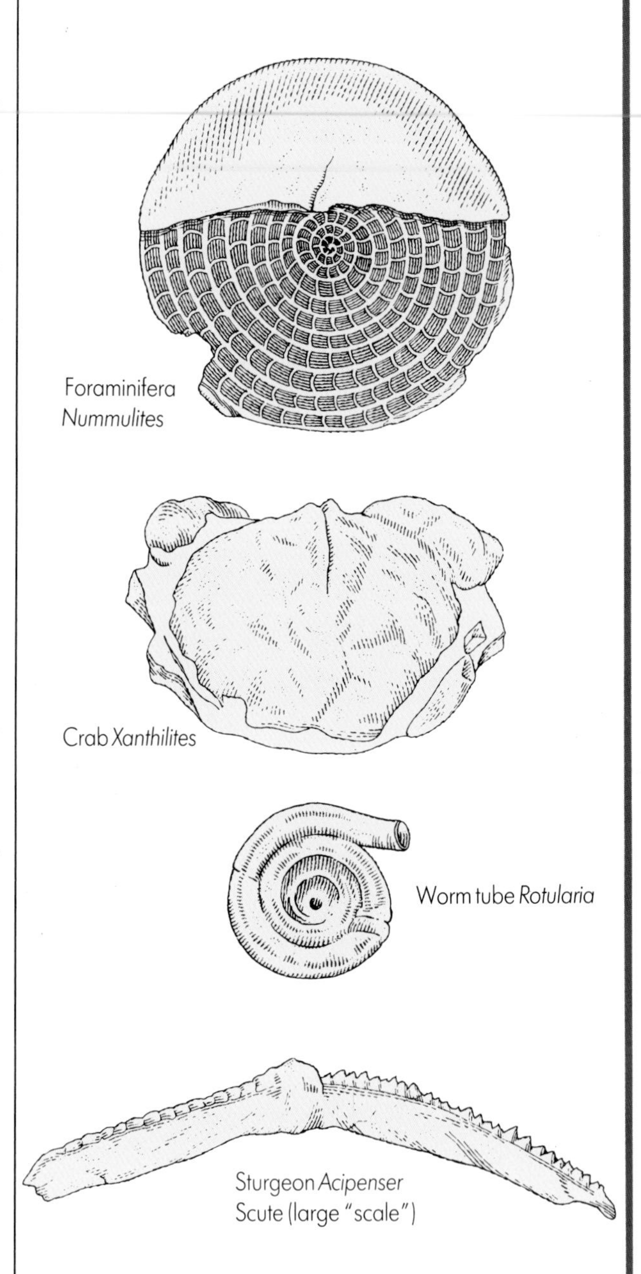

Foraminifera *Nummulites*

Crab *Xanthilites*

Worm tube *Rotularia*

Sturgeon *Acipenser* Scute (large "scale")

flowering plants, the grasses, during the latter half of the Tertiary. The development of a zonal climate during the Quaternary was quickly reflected by the zonal vegetation that stretches in bands around the Earth today.

Invertebrates

In the sea, foraminiferans again formed a major part of the marine plankton. Calcareous plankton repopulated the Paleocene seas, after being virtually wiped out at the end of the Mesozoic. These planktonic forms declined again during the Oligocene, to be replaced by modern forms in the Miocene. During the Pliocene, the poor climate reduced the variety of marine plankton.

The mass extinction that killed the dinosaurs also affected many other animals. Corals, bryozoans, brachiopods, echinoderms and crustaceans all lost representatives, but new ones soon replaced them. Many of the groups which had been so successful during the Mesozoic disappeared, including ammonites, belemnites and certain bivalves such as hipurites. Gastropods and bivalves like oysters, very similar to present-day species, took their place among the dominant invertebrates in the sea – indeed, some scallops learned to swim. Land gastropods (snails), first seen in the Carboniferous forests, flourished.

Among the arthropods, the Eocene saw further development of insects, particularly the bees, which lived on the pollen and nectar of flowering plants. There are many insect fossils from this period, particularly those preserved in amber. The insects were trapped while feeding on the sweet, sticky resin leaking from wounds on certain trees. These lumps of resin became fossilized when they fell into the muddy ooze beneath. Earwigs, flies, beetles, arachnids such as spiders, and even some small vertebrates like frogs and lizards, have been perfectly preserved in this way.

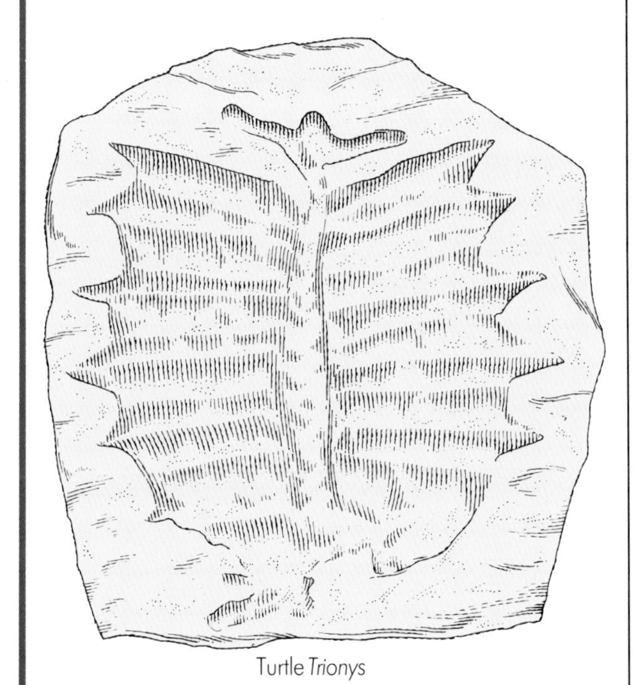

Turtle *Trionys*

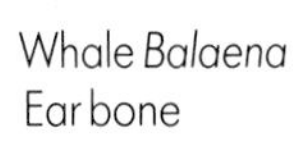

Whale *Balaena*
Ear bone

Snail *Cepaea*

Opposite page and above Tertiary and Quaternary fossils. Fossilized life-forms become more recognizable to us as evolution approaches present-day species. Nummulites are responsible for much of the deep-sea sediments which still form. Crabs, tube worms, the sturgeon, turtles, whales and the common snail *Cepaea* are all familiar animals today.

Vertebrates

The mass extinction that began the Cenozoic era had devastating effects on the vertebrates. The Age of Reptiles ended abruptly. The dinosaurs, at the height of their evolution with more species than ever before, were completely wiped out, along with other large reptiles of the sea and air, such as the mosasaurs, plesiosaurs, ichthyosaurs and pterosaurs. The only reptile groups to survive the extinctions still live on today: crocodilians, turtles, snakes, lizards and the tuatara. Among other vertebrates, there continued to be few amphibians, and these lived mainly in or near fresh water. The groups of bony fish that we know today such as cyprinids, trout, and bass developed in the Tertiary period. Sharks became more common. The fishes had the seas to themselves for a time.

By the beginning of the Cenozoic, birds had

(Continued on page 144.)

And so to the present

Fossils from the Cenozoic and Quaternary eras are closely related to present-day animals and plants. This was the age when more familiar mammals and flowering plants gained supremacy on land. Fossils become more easily recognizable to our untutored eyes, and can be more usefully compared with living forms than the more remote organisms from the distant past. Some of the plant and animal groups that evolved rapidly during this short period have already passed their peak of success. They are now extinct or nearly so, leaving only one or two endangered species, or stuffed specimens in museums, as clues to their past. The "snapshots" of the past, which we see when we excavate a fossil-rich bed, correspond to the "snapshot" of geological time which has passed since human history began.

(**The photographs shown here are slightly larger than life-size**.)

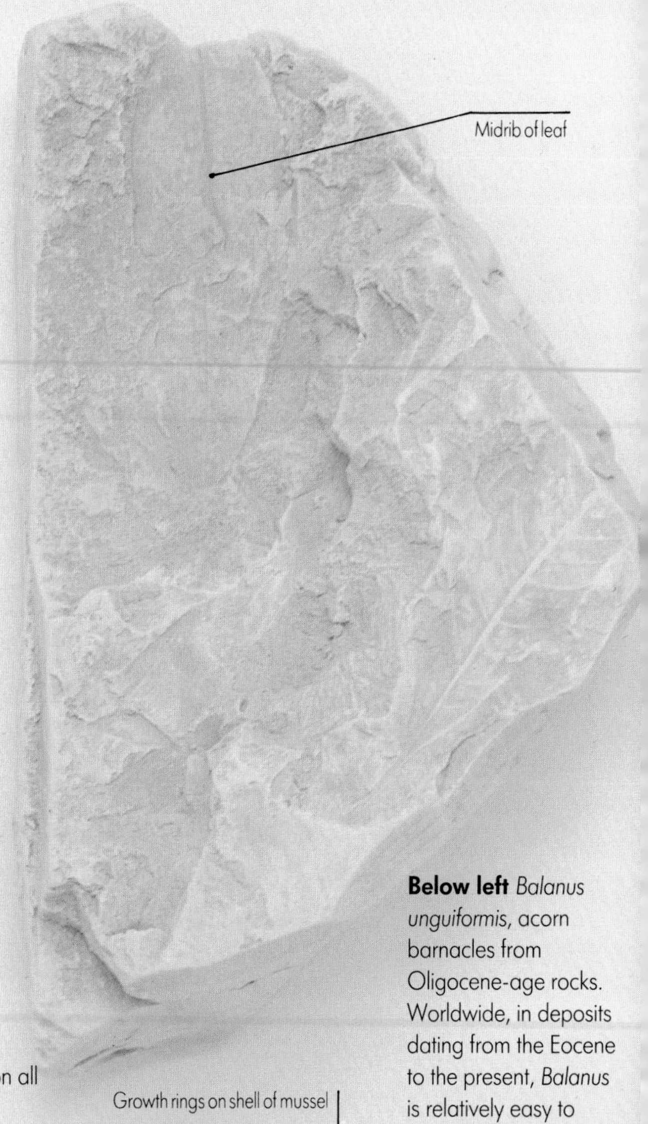

Below *Micraster coranguinum*, a sea urchin found in deposits from the Cretaceous to the Paleocene periods in Europe, Africa, Madagascar and Cuba. Their heart-shaped skeletons, or tests, are amenable to fossilization, and specimens from this group provide one of the best evolutionary sequences in paleontology.

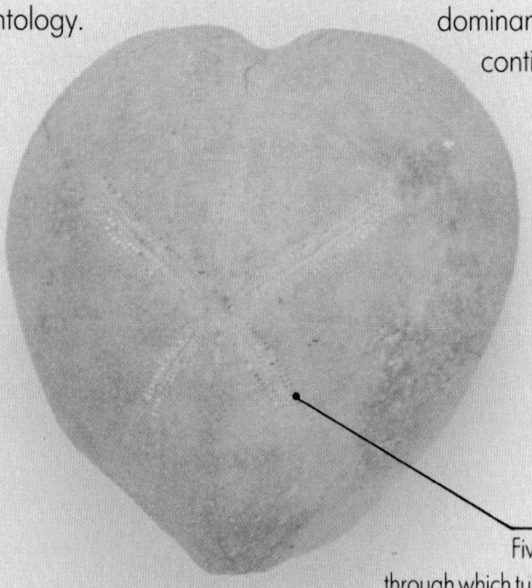

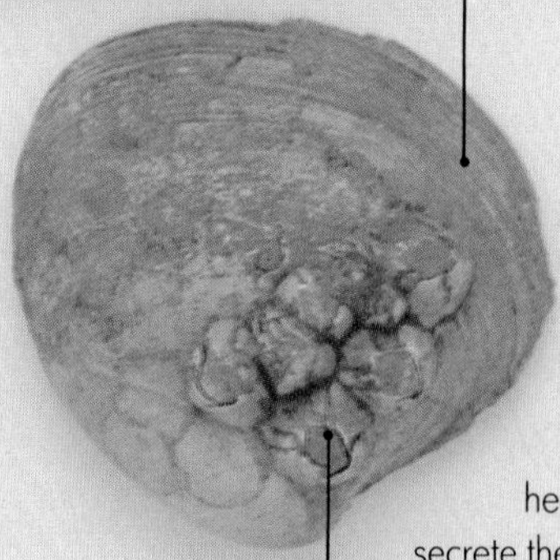

Right A laurel-type leaf from Eocene beds. These leaves first appear in sediments of about this age. They became more diverse and common in Oligocene and Miocene deposits. Flowering plants, of which laurels are only one of many hundreds of groups, first appeared during the Cretaceous period. They evolved rapidly and within 30 million years were the dominant plants on all continents.

Midrib of leaf

Growth rings on shell of mussel

Below left *Balanus unguiformis*, acorn barnacles from Oligocene-age rocks. Worldwide, in deposits dating from the Eocene to the present, *Balanus* is relatively easy to recognize. Its cone-shaped "skeleton" has six joined plates and is often found partly embedded in a substrate – in this case, probably a mussel. Barnacles are crustaceans, cousins of crabs. They fix themselves head-down to the substrate, secrete their plates around their soft bodies, and feed by kicking their legs through the water to strain out food.

Five double-rows of holes through which tube feet protruded in life

Barnacle

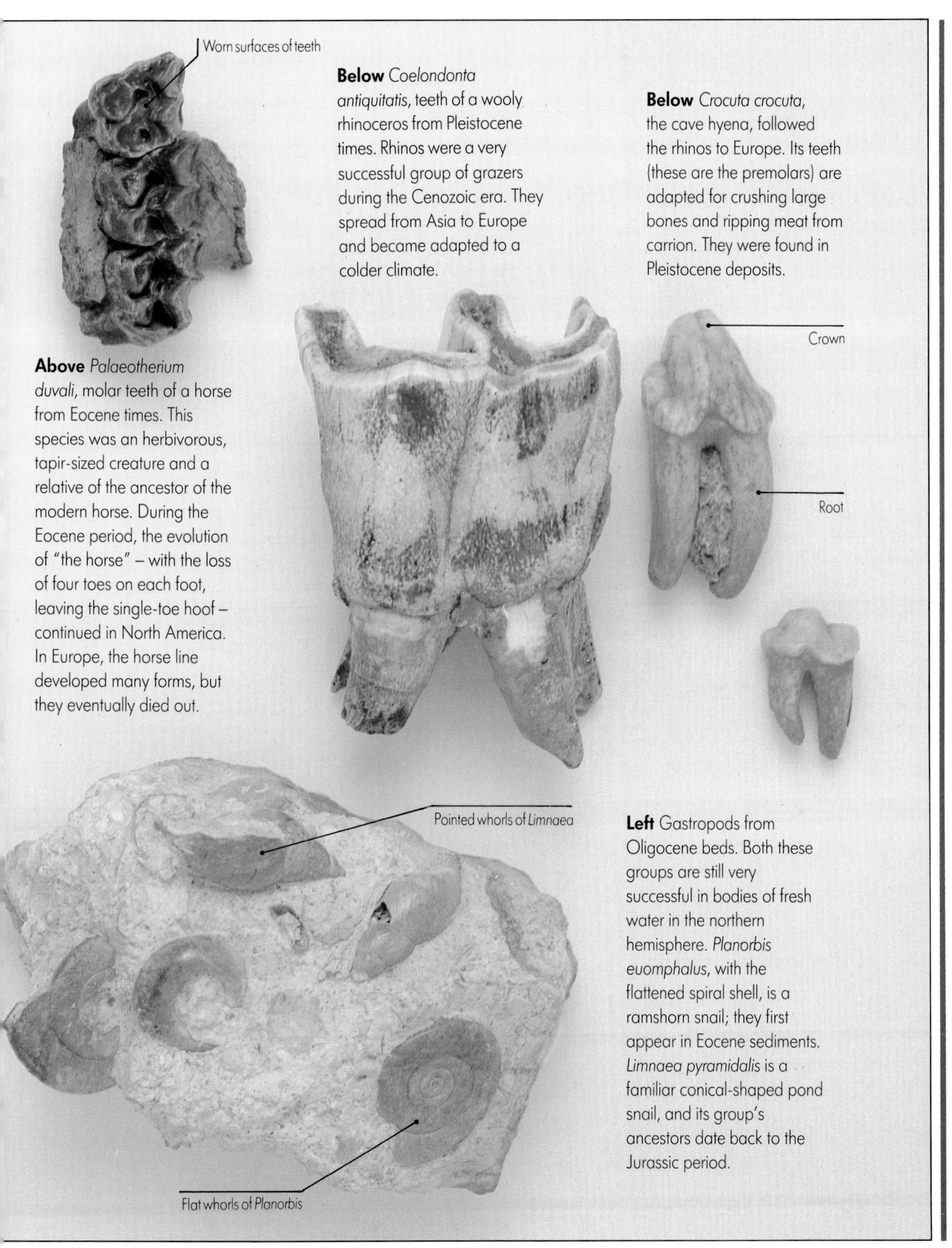

Above *Palaeotherium duvali*, molar teeth of a horse from Eocene times. This species was an herbivorous, tapir-sized creature and a relative of the ancestor of the modern horse. During the Eocene period, the evolution of "the horse" – with the loss of four toes on each foot, leaving the single-toe hoof – continued in North America. In Europe, the horse line developed many forms, but they eventually died out.

Below *Coelondonta antiquitatis*, teeth of a wooly rhinoceros from Pleistocene times. Rhinos were a very successful group of grazers during the Cenozoic era. They spread from Asia to Europe and became adapted to a colder climate.

Below *Crocuta crocuta*, the cave hyena, followed the rhinos to Europe. Its teeth (these are the premolars) are adapted for crushing large bones and ripping meat from carrion. They were found in Pleistocene deposits.

Left Gastropods from Oligocene beds. Both these groups are still very successful in bodies of fresh water in the northern hemisphere. *Planorbis euomphalus*, with the flattened spiral shell, is a ramshorn snail; they first appear in Eocene sediments. *Limnaea pyramidalis* is a familiar conical-shaped pond snail, and its group's ancestors date back to the Jurassic period.

already evolved into their main classes. After the dinosaurs, and before the mammals took hold, there were giant flightless birds that took the role of chief terrestrial predators. They reached their zenith in the Miocene, but representatives are now only found where placental mammals were late arrivals, in places such as South America and Australia. Early birds like *Odontopteryx* and *Lithornis* occurred in the Oligocene.

With the demise of the pterosaurs, the birds took over the skies. Some species also adapted to aquatic environments *(Hesperornis* of the Cretaceous was a gannet-like creature). Waders, pelicans, penguins and sea birds used the food found on and below the sea's surface and along its shores. Song birds, birds of prey, ducks and hummingbirds all evolved to fill niches left open by the demise of the flying reptiles and the coming of flowering plants.

Mammals

The Tertiary period is regarded as the "Age of Mammals." With most reptiles gone by the end of the Cretaceous, the land, air and sea were left rather sparsely populated. In fact, ancestors of all the main groups of mammals had already evolved by the Eocene. But they flourished and diversified, freed from reptilian domination, to populate the land and sea, and to fly at night, as bats, when most birds were inactive.

Paleocene mammals included ancestors of modern hoofed herbivores, the condylarths, as well as early rodents and squirrel-like primates. The Eocene saw more rodents, as well as bats, elephant ancestors, whales and artiodactyl and perissodactyl ungulates such as Eohippus, the "dawn horse." Several groups of more primitive mammals became extinct.

The appearance of grasses provided new ecological niches for mammalian grazers. Oligocene grazing mammals spread, as well as dogs, cats, pigs and rat-like rodents. Toothed whales replaced the early forms. By the Miocene, great herds of grazing animals roamed the grasslands. Among them were early horses, pigs, camels, rhinos, antelopes and a beaver-like animal that made huge corkscrew burrows. Fossil apes abounded in Eurasia and Africa.

During the Pliocene, the human line (hominids) appeared. The African savannas and their mammals evolved, and grasslands spread over much of the world.

In the Pleistocene epoch of the Quaternary, mammals moved south, away from the advancing ice. North American types, in particular, migrated across the new land link into South America, and the placentals began to replace the previously isolated marsupials. Some of Austrialia's marsupials reached giant proportions.

Creatures adapted to the ice-bound north. Reindeer, wooly mammoths and wooly rhinos lived there. They were soon to be hunted by a recently evolved human species that was spreading around the world.

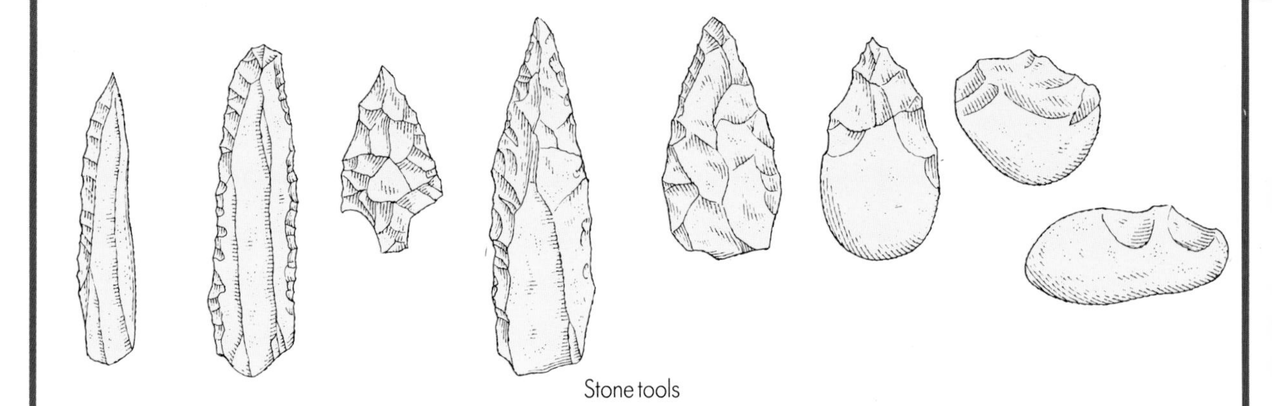

Stone tools

An uncomformity between desert sandstones of Permian times, about 250 millon years ago, and Pliocene gravels only a few million years old, at the Shoshone River in northern Wyoming. Two very different groups of fossils come from these deposits.

THE GAZETTEER

The following section lists some of the main fossil sites and museum displays to be found on each continent. To fuel your enthusiasm for fossil hunting, it is worth visiting these places to see how the experts locate, uncover and present their finds.

The most productive paleontological sites contain fossils of many kinds of plants and animals. The ways in which these are grouped together in layers give us great insight into which organisms lived at the same time, and the probable ecological relationships between them. Many sites continue to produce fossils as the excavations delve deeper into the rocks. Such places are rare and precious, and most countries treasure them. Access is restricted, to prevent damage to the remains not yet excavated.

On the other hand, people wishing to visit famous sites are a valuable source of income for funding further excavations, studies, and the preservation of the site. This is why some sites have tourist-type ticket offices and well-defined footpaths to the most famous spots.

However, do not be disappointed if the most interesting excavations, from the viewpoint of the keen amateur paleontologist, are hidden from public view. These productive sites are usually associated with a museum or university or similar institution, where the scientists involved are based. The museum may not necessarily be on site; it is often in the nearest large city. Here, the best finds from the site are displayed for public viewing. The general fascination for prehistoric life, particularly among children, guarantees a constant stream of visitors to the museums – again, a good source of income. Modern exhibition techniques can be innovative and exciting.

Large museums may operate a mailing list which publicizes forthcoming special exhibitions, opening times, entrance fees and other details. There may be a newsletter that reports on recent acquisitions, controversies and discoveries at the various associated sites. Write to your local museum, as well as the main city museum in your area, explaining your interests and asking what is available.

Left A fine mounted stegosaurian dinosaur in Zigong Dinosaur Museum, Sichuan Province, China. This specimen is one of the Taibaihuyayang dinosaurs, unearthed around 1981. The creature lived about 150 million years ago and is a relative of the more famous *Stegosaurus* of North America.

Above China's Sichuan Province has yielded many fossils. Here, workers clean and prepare the remains of a giant sauropod dinosaur.

AFRICA

Africa is now known to be the "Cradle of Humankind." The Olduvai Gorge site in northern Tanzania has yielded many early hominid fossils, as have the areas of Lake Turkana in Kenya, Hadar in Ethiopia and South Africa. Most of what is known about early human evolution is based on the study of these fossils.

Once part of the supercontinent of Gondwanaland, Africa has remained stable for much of its long history and has been at the center of the evolution of land plants and animals. At times, much of the land was under shallow seas or ice fields, so fossil finds also include those of fish and marine reptiles.

There are many African examples of the mammal-like reptiles that dominated vertebrate evolution before the rise of the dinosaurs. Dinosaur fossils are also well represented throughout the continent, and the great grasslands have influenced the evolution of grazing vertebrates.

Museums to visit

Cape Town, South Africa
South African Museum
Prehuman remains, many mammal fossils

Harare, Zimbabwe
National Museum of Zimbabwe
General fossils and geology

Johannesburg, South Africa
Bernard Price Institute of Palaeontology
General fossils and geology

Nairobi, Kenya
Kenya National Museum
Many prehuman remains

Niamey, Niger
National Museum
General fossils and geology

Rabat, Morocco
Museum of Earth Sciences
General fossils and geology

Famous fossil sites

Republic of Mali, in the Tilemsi Valley in the Sahel region south of the Sahara, the bare, rocky and sandy terrain is littered in some places with the fossil remains of fish, crocodiles and turtles – evidence of a north-south seaway through this part of Africa.

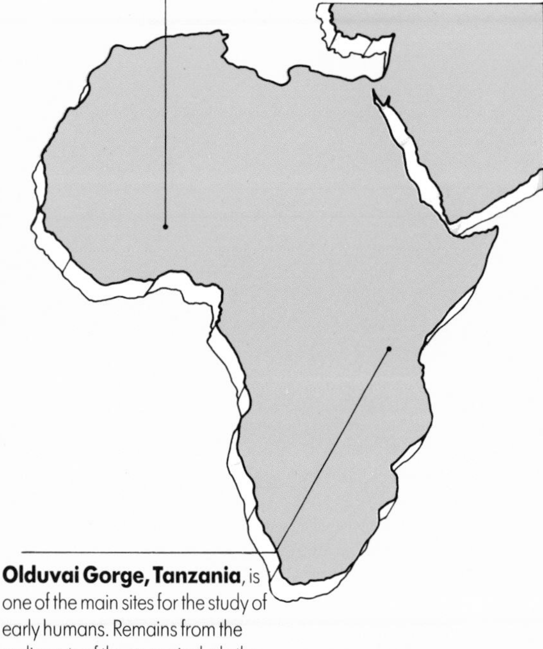

Olduvai Gorge, Tanzania, is one of the main sites for the study of early humans. Remains from the sediments of the gorge include the first ever discovery of early hominid fossils, dated at 1.8 million years, found with stone tools. This find was discovered in 1959 by Dr. Mary Leakey.

A restoration of a mammal-like reptile, a dinocephalian, in the Late Permian landscape of the Karroo, South Africa. The display is in the South African Museum, Cape Town.

NORTH AND SOUTH AMERICA

The American continent is composed of rocks of all ages, which contain a corresponding variety of fossils. Plants, invertebrates, fish, reptiles, birds and mammals are all represented. The famous Burgess Shales of Canada have produced thousands of examples of some of the earliest fossils, while Arizona boasts the largest petrified forest in the world. Recent mammals are recorded in the Rancho la Brea tar pits of Los Angeles.

The large areas of Mesozoic sedimentary rocks in central North America were the site of the famous exploits of Marsh and Cope at the end of the last century (page 20). They did much to advance the science of paleontology, and large numbers of dinosaur fossils are still being found in these rocks, as well as in Canada and South America. Indeed, dinosaurs are represented from their earliest beginnings to their ultimate extinction.

Famous fossil sites

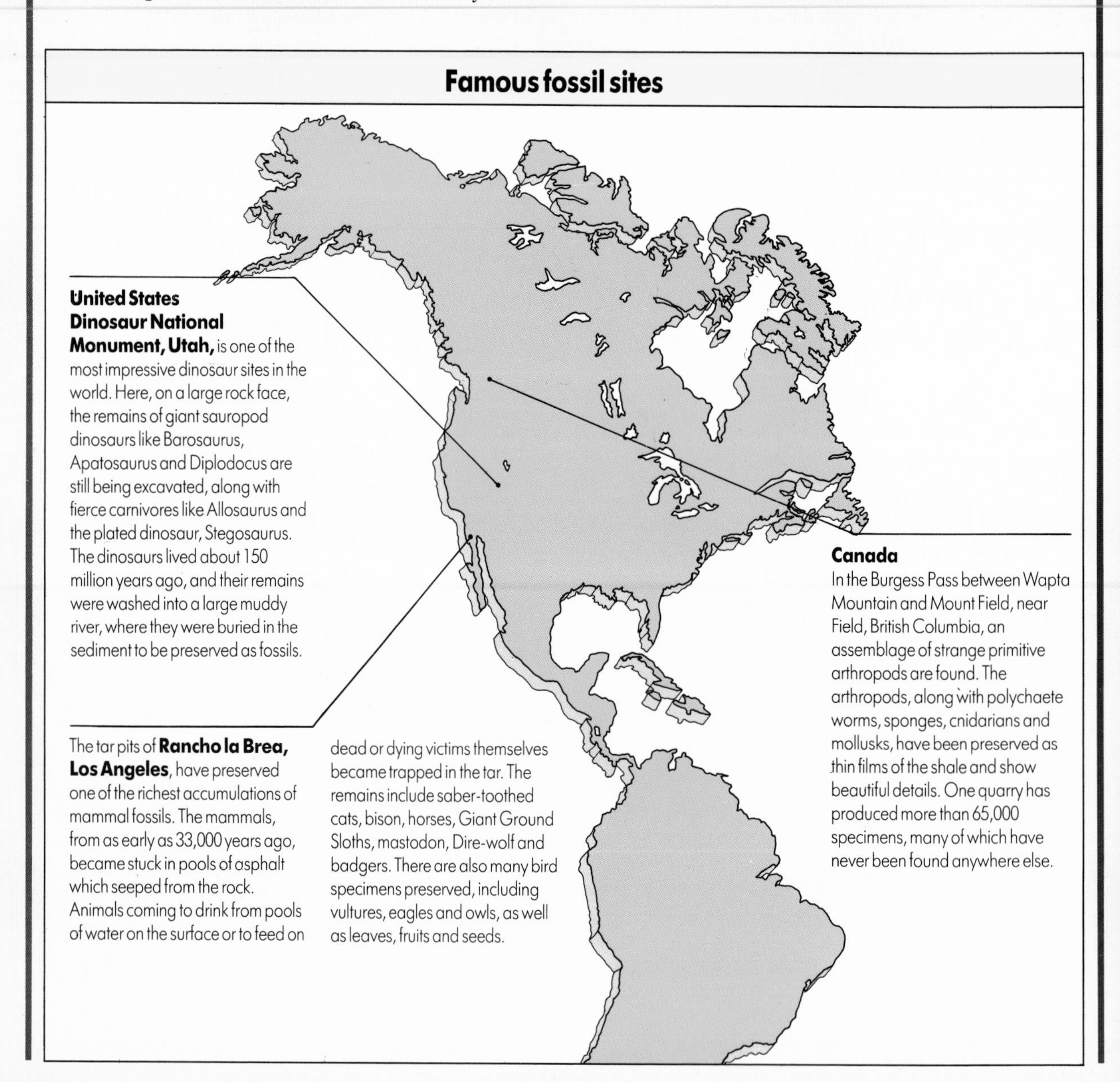

United States
Dinosaur National Monument, Utah, is one of the most impressive dinosaur sites in the world. Here, on a large rock face, the remains of giant sauropod dinosaurs like Barosaurus, Apatosaurus and Diplodocus are still being excavated, along with fierce carnivores like Allosaurus and the plated dinosaur, Stegosaurus. The dinosaurs lived about 150 million years ago, and their remains were washed into a large muddy river, where they were buried in the sediment to be preserved as fossils.

The tar pits of **Rancho la Brea, Los Angeles**, have preserved one of the richest accumulations of mammal fossils. The mammals, from as early as 33,000 years ago, became stuck in pools of asphalt which seeped from the rock. Animals coming to drink from pools of water on the surface or to feed on dead or dying victims themselves became trapped in the tar. The remains include saber-toothed cats, bison, horses, Giant Ground Sloths, mastodon, Dire-wolf and badgers. There are also many bird specimens preserved, including vultures, eagles and owls, as well as leaves, fruits and seeds.

Canada
In the Burgess Pass between Wapta Mountain and Mount Field, near Field, British Columbia, an assemblage of strange primitive arthropods are found. The arthropods, along with polychaete worms, sponges, cnidarians and mollusks, have been preserved as thin films of the shale and show beautiful details. One quarry has produced more than 65,000 specimens, many of which have never been found anywhere else.

Redpath Museum, Montreal, Quebec. Museum displays usually concentrate on interesting finds from the local area. The displays should give plenty of information about fossils and their formation, and the types of rocks likely to reveal them. A good exhibition also shows how the specimens fit into the general picture of the history of life on Earth, and how the original organisms may have lived their lives. Note the bilingual text here, in French and English.

Museums to visit

North America

Austin, Texas
Texas Memorial Museum
Early reptiles and mammals

Berkeley, California
University of California Museum of Paleontology
Mesozoic reptiles

Cambridge, Massachusetts
Museum of Comparative Zoology, Harvard University
Fossil vertebrates

Chicago, Illinois
Field Museum of Natural History
Fossil plants, invertebrates and vertebrates

Cleveland, Ohio
Natural History Museum
Fossil fish, dinosaurs and mammals

Los Angeles, California
Los Angeles County Museum of Natural History
Fossils from tar pits (Rancho la Brea)

New Haven, Connecticut
Peabody Museum of Natural History
Fossil vertebrates

New York City, New York
American Museum of Natural History
Fossil vertebrates, including largest collection of dinosaurs in the world

Denver, Colorado
Denver Museum of Natural History
Fossil reptiles and mammals

Jensen, Utah
National Dinosaur Museum
Dinosaur fossil beds, excavation and cleaning laboratories

Montreal, Quebec, Canada
Redpath Museum
Various fossils

Pittsburgh, Pennsylvania
Carnegie Museum of Natural History
Major fossil collections

Toronto, Ontario, Canada
Royal Ontario Museum
Vertebrate fossils

Washington D.C.
National Museum of Natural History
Major fossil collections

South America

Buenos Aires, Argentina
Museo Argentino de Ciencias Naturales
Dinosaurs, other fossils, giant mammals

Mexico City, Mexico
Natural History Museum, Mexico City
Various fossils

Rio de Janeiro, Brazil
Museo Nacional, Rio de Janeiro
Dinosaurs and other fossils

ASIA

The geology of Asia is dominated by several periods of mountain building. The Himalaya Mountains are the youngest mountains in the world, and fossils of sea invertebrates found on their slopes prove that the sediments once lay beneath the sea. Asia sits on tectonic plates which moved together after the break-up of Pangaea and Gondwanaland. Most of the region is still vulnerable to earth movements today.

Many museums contain dinosaur fossils, as well as plant, invertebrate and other vertebrate fossils. The Gobi Desert has yielded many dinosaur remains, including nests and eggs. Both China and India also boast impressive dinosaur finds, while in Siberia, frozen mammoth remains are still being found by the ton!

Museums to visit

Beipei, China
Beipei Museum
General fossils and geology

Calcutta, India
Indian Statistical Institute
Geology, surveying, fossils

Osaka, Japan
Museum of Natural History
General fossils and geology

Peking, China
Institute of Vertebrate Paleontology; fine dinosaurs and general collections

Sichuan Province, China
Zigong Dinosaur Museum
Complete dinosaur fossils

Tokyo, Japan
National Science Museum
General fossils and geology

Ulan Bator, Mongolia
Mongolian Academy of Sciences; dinosaurs and Gobi Desert fossils

Famous fossil sites

Russia
In the frozen lands of northern Siberia, the remains of fossil mammals are found. The mammoths were so numerous that it is estimated that 55,000 tons of fossil ivory tusks are still buried offshore along the 600-mile (100- km) long coast between the Rivers Yana and Kolyma. For many centuries, there has been an important trade in the fossil ivory, used for ornaments and jewelry.

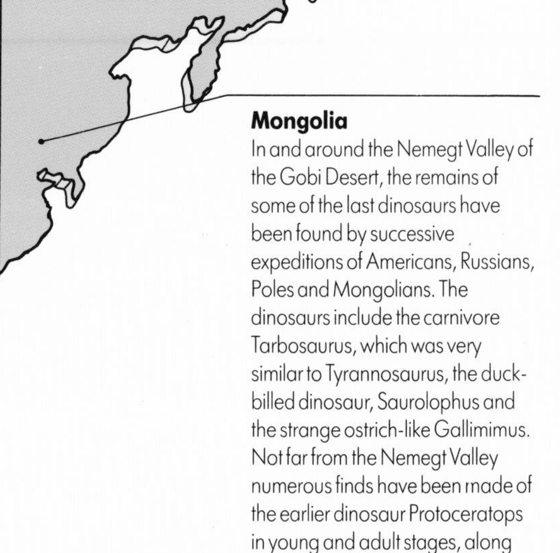

Mongolia
In and around the Nemegt Valley of the Gobi Desert, the remains of some of the last dinosaurs have been found by successive expeditions of Americans, Russians, Poles and Mongolians. The dinosaurs include the carnivore Tarbosaurus, which was very similar to Tyrannosaurus, the duck-billed dinosaur, Saurolophus and the strange ostrich-like Gallimimus. Not far from the Nemegt Valley numerous finds have been made of the earlier dinosaur Protoceratops in young and adult stages, along with many finds of fossilized eggs.

Large numbers of dinosaurs have been found during this century in China. This specimen of *Gazosaurus*, a swift-running predatory species, is mounted and displayed at the Zigong Dinosaur Museum, Sichuan.

AUSTRALIA

Australia has been largely isolated from other major landmasses since it split from Gondwanaland, 160 million years ago. The continent is based on very ancient Precambrian rocks, among the oldest so far found in the world. Traces of the earliest life forms have been found in these rocks. Parts of the land surface have been flooded from time to time, and sediments contain remains of sea creatures. The most impressive of these are the beautiful Ediacara fossils of Precambrian soft-bodied animals, from the Ediacara Hills of the Flinders Range, South Australia. There is also a large collection of graptolites from Victoria.

Fossils of all main groups of animals and plants have been found in Australia. Devonian lungfish come from the northwest, and dinosaur remains appear throughout the continent. The relatively recent evolution of mammals in Australia reflects the isolation of the continent. Placental mammals never developed there, and suitable ecological niches are (or at least were) filled by marsupial mammals. The large flightless birds which dominated the land between the Age of Reptiles and the Age of Mammals are also still represented on this continent, both by fossils and by the living emu and cassowary.

Museums to visit

Adelaide, South Australia
South Australian Museum
General fossils and geology

Brisbane, Queensland
Queensland Museum
General fossils and geology

Melbourne, Victoria
Geological Museum
General fossils and geology

Sydney, New South Wales
Australian Museum
Main collections in Australia, including giant extinct marsupials

Famous fossil sites

At **Gogo** in northwestern Australia, nodules of limestone are found containing the fossils of lung fish from 400 million years ago. The fish are perfectly preserved with all their hard parts remaining, including the scales. Acids, which dissolve the rock and leave the bony parts intact, are often used to extract these fossils from the rock.

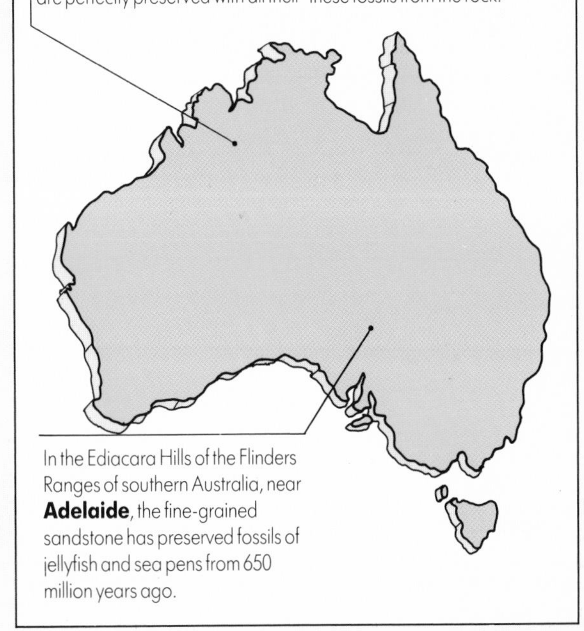

In the Ediacara Hills of the Flinders Ranges of southern Australia, near **Adelaide**, the fine-grained sandstone has preserved fossils of jellyfish and sea pens from 650 million years ago.

Dinosaur exhibits are a crowd-puller at the Museum of Victoria in Melbourne, Australia.

EUROPE

The predominance of fossil finds, sites and collections in Europe is due more to the large numbers of people looking for fossils than to the richness of fossiliferous rocks. It was mainly in Europe, in the last two to three centuries, that the science of paleontology developed. Many of the most extensive collections were begun in Victorian times.

As a consequence, there are impressive arrays of plants, invertebrates and vertebrates in even the smallest museum. There are many large collections of invertebrates such as graptolites and trilobites, as well as petrified forests and fossilized giant "sea scorpions." The evolution of fish, reptiles, birds and mammals is studied at the many centers of learning throughout Europe. Mammal remains are also common in the tillate left by the ice which covered most of Europe until recent times.

Famous fossil sites

Baltic
Around Samland Promontory, amber is abundant as it eroded from the sediments deposited some 50–55 million years ago. The amber was fossilized from the resin of the extinct pine tree, *Pinus succinifera*, and is remarkable for the preservation of insects which were trapped in the once sticky resin.

Scotland
Dobb's Linn, near Moffat in the Southern Uplands, contains one of the classic Ordovician localities in the world. The dominant fossils are graptolites, and their changing succession through the rocks shows the change from Ordovician rocks into Silurian rocks some 460–430 million years ago.

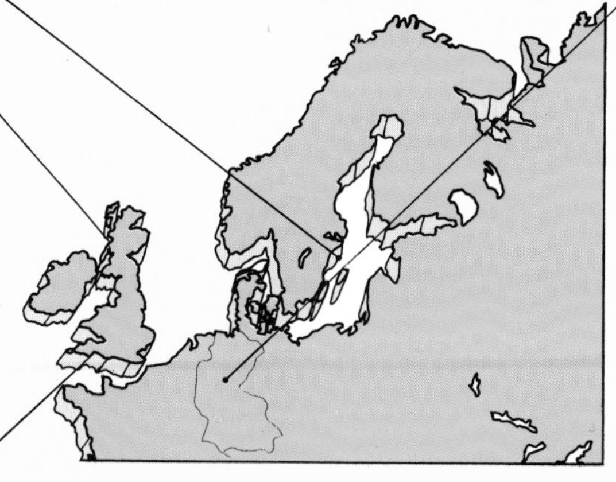

West Germany
At Bundenbach and in an area south of the River Mosel, the Hunsruck Slate contains wonderful invertebrate fossils preserved in the mineral iron pyrite. The fossils are about 390 million years old and include trilobites, starfish and crinoids. The fossils, although flattened, preserve many details: by using X-rays, details such as ink sacs of belemnites and eye structures of trilobites can be seen. The rocks are of the Early Devonian era, about 390 million years old.

In the Solnhofen Limestone of Bavaria, the fine-grained rock has preserved the finest details of crayfish, birds, fish, pterodactyls, crabs, insects, dinosaurs and the famous Archaeopteryx.

England
Lyme Regis, Dorset, has been known since the 17th century as a good locality for vertebrate fossils such as ichthyosaurs. The fossils are found in hard Jurassic limestone deposited about 190 million years ago. One of the most famous fossil collectors at Lyme Regis was Mary Anning, who is reputed to have found her first ichthyosaur specimen in 1810 when she was 11 years old.

Part of the huge collection of fossil vertebrates displayed in the main hall of the Bavarian State Institute for Paleontology and Historical Geology, Munich. Note the Mesozoic pterosaur flying over the much more recent elephant!

Museums to visit

Athens, Greece
Department of Geology and Paleontology; general fossils

Berlin, Germany
Natural History Museum, Humboldt University
Has world's largest mounted dinosaur

Brussels, Belgium
Royal Institute of Natural Sciences
Dinosaurs, world's largest collection of *Iguanodon* skeletons

Cardiff, Wales
National Museum of Wales
Cambrian fossils

Edinburgh, Scotland
Royal Scottish Museum
Devonian Old Red Sandstone exhibits

London, England
British Museum (Natural History)
One of the world's premier fossil collections

Munich, Germany
Bavarian State Institute for Paleontology and Historical Geography; general fossils

Paris, France
National Museum of Natural History
Largest fossil collection in France

Prague, Czechoslovakia
Narodoni Museum
General fossil collections

Rome, Italy
Museum of Paleontology
General fossils and geology

Vienna, Austria
Vienna Natural History Museum
Dinosaurs and ice-age mammals

Useful Addresses

If you wish to find out more about fossils and fossil hunting, there are several methods you could try:

- You might visit the main library in the nearest town or city. Ask to see their list of organizations and clubs in the area. There is sometimes a club devoted to paleontology, or at least one for people interested in archaeology or geology, which may have a special group interested particularly in fossils. Contact the membership secretary for further information.

- The major museums mentioned on the previous pages should also be able to help, by putting you in touch with appropriate organizations in your region. Write to the curator of the paleontology or geology department.

- Universities are another place for information. Again, write to the head of the paleontology or geology department.

- Look in the yellow pages, for example, for entries under Paleontology, Geology, Geological Surveys, Museums, Academic Institutions, Mining and Prospecting and so on.

- You may find advertisements for museums, fossil collections, "dinosaur parks" etc., in tourist or publicity brochures and books produced by the state tourist board, or by the regional tourist office.

- In North America: Paleontology Society, Inc.; c/o New Mexico Bureau of Mines; Socorro, New Mexico 87801.
Earth Watch; 680 Mt Auburn Street; Box 403; Watertown, Massachusetts, 02172.

- In South America: Paleontological Association of Argentina; Maipu 645, ler Piso; 1006 Buenos Aires, Argentina.

- In the U.K.: Palaeontological Association London; c/o The Secretary; The Croft Barn; Church Street; East Hendred, Oxon OX12 8LA, England.

- In Australia: The Fossil Collectors Association of Australia; 15 Kenbury Road; Heathmont, Victoria 3135, Australia.

- In Asia: Geological Survey of India; 29 Jawaharlal Nehru Road; Calcutta 700016, India.

INDEX

Page numbers in *italic* refer to the illustrations.

INDEX

ACKNOWLEDGMENTS

Quarto and the author would like to thank the following for their help with this publication. Every effort has been made to obtain copyright clearance, and we do apologize if any omissions have been made.

p 9 GSF Picture Library; **p 10** GSF Picture Library; **p 13** GSF Picture Library; **p 14** Ann Ronan Picture Library; **p 15** above: Royal Library, Copenhagen/below: University of Bristol, Department of Geology; **p 16** left: Imitor/right above: Ann Ronan Picture Library/below: University of Bristol, Department of Geology; **p 17** above: Ann Ronan Picture Library/below: Bulloz; **p 18** above: Ann Ronan Picture Library/below: University of Bristol, Department of Geology; **p 19** left: Hulton-Deutsch Collection/centre: Ann Ronan Picture Library/right: Ann Ronan Picture Library; **p 20** Hulton-Deutsch Collection; **p 21** above: Los Angeles County Museum/below: Ann Ronan Picture Library; **p 22** The Telegraph Colour Library; **p 23** above: GSF Picture Library/below left: Dr A R I Cruickshank/ below right: Popperfoto; **p 24** Ann Ronan Picture Library; **p 25** GSF Picture Library; **p 26** GSF Picture Library; **p 27** below left: University of Bristol, Department of Geology/ center: GSF Picture Library/right: GSF Picture Library; **p 28** above: Dr A R I Cruickshank/below: GSF Picture Library; **p 29** above: University of Bristol, Department of Geology/below: GSF Picture Library; **p 30** above: GSF Picture Library/below left: Dr A R I Cruickshank/below right: GSF Picture Library; **p 32** University of Bristol, Department of Geology; **p 33** above left: GSF Picture Library/center left: GSF Picture Library/below left: University of Bristol, Department of Geology/below right: Dr A R I Cruickshank; **p 34** GSF Picture Library; **p 35** University of Bristol, Department of Geology; **p 36** University of Bristol, Department of Geology; **p 37** above: University of Bristol, Department of Geology/below: Secol, Thetford; **p 38** University of Bristol, Department of Geology; **p 39** University of Bristol, Department of Geology; **p 40** GSF Picture Library; **p 41** GSF Picture Library; **p 42** Quarto; **p 43** Quarto; **p 44** GSF Picture Library; **pp 46-7** Frank Lane Picture Agency; **p 47** above right: Frank Lane Picture Agency/below right: Derek Widdicombe; **p 48** Tyrrell Museum, photo Collin Orthner; **p 49** above: GSF Picture Library/below: Dr A R I Cruickshank **p 51** above: Derek Widdicombe/below: Dr M A Taylor; **p 53** Dinosaur Provincial Park; **p 54** Tyrrell Museum, photo Collin Orthner; **p 55** above: Dr M A Taylor/below: Dr A R I Cruickshank; **p 56** above: Billie Love/below: Steve Parker; **p 57** above: University of Bristol, Department of Geology/below: Steve Parker; **pp 58-9** GSF Picture Library; **p 60** British Museum Natural History/Quarto; **p 62** Dr Michael Benton; **p 65** Dr A R I Cruickshank; **p 66** above: GSF Picture Library/ below left: Dr M A Taylor/below right: Tyrrell Museum; **p 67** University of Bristol, Department of Geology; **p 69** GSF Picture Library; **p 70** GSF Picture Library; **p 72** GSF Picture Library; **p 74** Dr Michael Benton; **p 75** above: GSF Picture Library; **p 76** Quarto; **p 77** left: GSF Picture Library/right: University of Bristol, Department of Geology; **p 78** University of Bristol, Department of Geology; **p 79** above: University of Bristol, Department of Geology/center: GSF Picture Library/below: GSF Picture Library; **p 80** Quarto; **p 81** Quarto; **p 83** above left: Ann Ronan Picture Library/above right: University of Bristol, Department of Geology/below: J-L Charmet; **p 84** above: University of Bristol, Department of Geology/centre: GSF Picture Library/below left: University of Bristol, Department of Geology/below right: University of Bristol, Department of Geology; **p 86** University of Bristol, Department of Geology; **p 87** GSF Picture Library; **p 88** Frank Lane Picture Agency; **p 90** British Museum/Natural History; **p 91** Professor H Rieber; **p 92** above: Dr A R I Cruickshank/below: GSF Picture Library; **p 94** Dr Michael Benton; **p 95** Quarto; **p 96** The Telegraph Colour Library; **p 97** Dr M A Taylor; **p 98** University of Bristol, Department of Geology; **p 99** Glasgow Parks and Recreation; **p 100** GSF Picture Library; **p 101** University of Bristol, Department of Geology; **p 103** Ann Ronan Picture Library; **p 104** University of Bristol, Department of Geology; **p 106** Popperfoto; **p 109** Dr Michael Benton; **p 115** GSF Picture Library; **p 119** University of Bristol, Department of Geology; **p 123** Dr A R I Cruickshank; **pp 126-7** University of Bristol, Department of Geology; **p 131** GSF Picture Library; **pp 134-5** University of Bristol, Department of Geology; **pp 138-9** GSF Picture Library; **pp 142-3** University of Bristol, Department of Geology; **p 145** GSF Picture Library; **p 146** Xinhua News Agency; **p 147** Xinhua News Agency; **p 149** Dr Gillian King; **xp 151** Redpath Museum; **p 152** Xinhua News Agency; **p 153** Thomas Rich, Museum of Victoria; **p 155** Dr F Westphal.